Management and Industrial Engineering

Series Editor

J. Paulo Davim, Department of Mechanical Engineering, University of Aveiro, Aveiro, Portugal

This series fosters information exchange and discussion on management and industrial engineering and related aspects, namely global management, organizational development and change, strategic management, lean production, performance management, production management, quality engineering, maintenance management, productivity improvement, materials management, human resource management, workforce behavior, innovation and change, technological and organizational flexibility, self-directed work teams, knowledge management, organizational learning, learning organizations, entrepreneurship, sustainable management, etc. The series provides discussion and the exchange of information on principles, strategies, models, techniques, methodologies and applications of management and industrial engineering in the field of the different types of organizational activities. It aims to communicate the latest developments and thinking in what concerns the latest research activity relating to new organizational challenges and changes world-wide. Contributions to this book series are welcome on all subjects related with management and industrial engineering. To submit a proposal or request further information, please contact Professor J. Paulo Davim, Book Series Editor, pdavim@ua.pt.

More information about this series at http://www.springer.com/series/11690

Kaushik Kumar · J. Paulo Davim
Editors

Supply Chain Intelligence

Application and Optimization

Springer

Editors
Kaushik Kumar
Department of Mechanical Engineering
Birla Institute of Technology, Mesra
Ranchi, Jharkhand, India

J. Paulo Davim
Department of Mechanical Engineering
University of Aveiro
Aveiro, Portugal

ISSN 2365-0532 ISSN 2365-0540 (electronic)
Management and Industrial Engineering
ISBN 978-3-030-46424-0 ISBN 978-3-030-46425-7 (eBook)
https://doi.org/10.1007/978-3-030-46425-7

This Springer imprint is published by the registered company Springer Nature Switzerland AG
The registered company address is: Gewerbestrasse 11, 6330 Cham, Switzerland

Preface

The editors are pleased to present the book *Supply Chain Intelligence—Application and Optimization* under the book series *Management and Industrial Engineering.* The book title was chosen as it depicts upcoming trends in the management of industrial world for the next decade. This book is a compilation of different aspects of the same.

In the present period of the 4th Industrial Revolution "Industry 4.0" (I40) which calls for analog to digital transformation of the entire industrial production utilizing an amalgamation of manufacturing sector with Internet and Information and Communication Technologies (ICT). All the focus now has been primarily directed towards customer satisfaction, which calls for optimal cost, time, and quality. Aimed toward these three variables, companies are striving continuously to develop/improve practices and techniques to fulfil consumer requirements and, in turn, increase their market share and profit.

Hines (2004: p76) provided a customer-focused definition of *supply chain* stating "*Supply chain strategies require a total system view of the links in the chain that work together efficiently to create customer satisfaction at the end point of delivery to the consumer. As a consequence, costs must be lowered throughout the chain by driving out unnecessary expenses, movements, and handling. The main focus is turned to efficiency and added value, or the end user's perception of value. Efficiency must be increased, and bottlenecks removed. The measurement of performance focuses on total system efficiency and the equitable monetary reward distribution to those within the supply chain. The supply-chain system must be responsive to customer requirements*". In recent decades, globalization, outsourcing, and information technology have enabled many organizations to successfully operate collaborative supply networks in which each specialized business partner focuses on only a few key strategic activities.

In the twenty-first century, changes in the business environment have contributed to the development of supply chain networks. First, as an outcome of globalization and the proliferation of multinational companies, joint ventures, strategic alliances, and business partnerships, significant success factors were identified, complementing the earlier "just-in-time", lean manufacturing, and agile manufacturing practices. Second, technological changes, particularly the dramatic fall in communication costs (a significant component of transaction costs), have led to changes in coordination among the members of the supply chain network. Many researchers have recognized supply network structures as a new organizational form, using terms such as "Keiretsu", "Extended Enterprise", "Virtual Corporation", "Global Production Network", and "Next-Generation Manufacturing System". In general, such a structure can be defined as "a group of semi-independent organizations, each with their capabilities, which collaborate in ever-changing constellations to serve one or more markets in order to achieve some business goal specific to that collaboration".

Manufacturing sectors till the last decade, for profit and survival, completely dedicated their energy and expertise towards mass manufacturing for most of their work. But with the evolution of new era of digitalization and customization, industries are changing their focus from mass to customized manufacturing and creating newer concepts of effective supply chain activities to stay in the market. Now buzz words like *Strategic Sourcing Decisions, Supply Chain Modeling, Inventory Modeling, Cold Chain, Block Chain, Logistics Modeling, Supply Chain and Big Data, Evolutionary Optimization, Hybrid Optimization* has been incorporated in the supply chain concept.

The main objective of the book is dedicated to Micro- and Nanomachining and is targeted to cater to the needs of all Academics Students, Researchers, and Industry Practitioners, Engineers, Research Scientists/Academicians involved in machining at micro/nanolevel towards creating Macro- or Micro-sized components.

In view of the changing scenario, this book has sections providing general introduction and applications of the newer concepts of *Supply Chain Management* (SCM). This interdisciplinary book will provide a bridge between management researchers/decision-makers and computer analysts and fills this gap by providing state-of-the-art descriptions of the corresponding problems and advanced methods for solving them for industrial fraternity.

The chapters in the book have been categorized into four parts, namely *Supply Chain*; *Cold Chain*; *Block Chain*; and *Applications*.

The first part contains the first three chapters, whereas the second part has fourth and fifth chapter; the third part with sixth and the fourth part contains seventh and eight chapter.

The first part starts with chapter "An Overview of Supply Chain Dynamics from a Behavioral Operations Perspective" which provides an overview of the state of the art in Behavioral operations management studies focusing on supply chain dynamics. The chapter provides insight on how, in the last decade, this field of study has been gaining momentum in scholarly research. Additionally, it also focuses on supply chain dynamics and perform a systematic literature review on the behavioral causes of the bullwhip effect. Finally, it offers an illustration of how multi-method research can enrich behavioral operations management by presenting the results of a study in which a behavioral cause of supply chain dynamics, namely risk aversion, is explored by combining human experiments and simulation.

Chapter "Classification of Factors Associated with a Closed-Loop Supply Chain System, Their Modelling Methods and Strategies" provides an insight to environmental impact of the industries due to nonsustainability in their open-end supply chain systems and elaborates on closed-loop supply chain system (CLSC). It is an improved system often suggested by researchers to enhance sustainability and lower environmental degradation by closing the loop of the product lifecycle and encouraging product returns and other green activities. The concept was strengthened by the introduction of various policies and regulations by the government for the sake of maintaining the health of the environment. In this chapter, a thorough investigation of the CLSC system and categories/groups of factors influencing the optimal parameters of a CLSC system have been reported including modeling of CLSC strategies involving different uncertainty problems which are required to be solved giving an account of various modeling methods and solution strategies commonly used in the research of CLSC systems for solving the uncertainty problems and analyze different aspects of the practical system for an optimized supply chain system.

Chapter "Project-Based Supply Chain Intelligence and Digital Fabrication for a Sustainable Film Industry", the last chapter of Section I, enlightens the readers with societal, political, and legislative drive towards greater transparency and accountability for environmental, social, and economic sustainability in business recognizing the complexities involved in monitoring operations in manufacturing and construction industries, and in particular where those are project-based. The chapter provides a solution to the above by developing digital technology to provide new ways of tracking elements within supply chain operations, as well as new communication platforms, fabrication techniques, and data analysis capabilities. This chapter provides an argument for the need for transition research in project-based manufacturing and construction industries, with film production as an example, to provide the conceptualization of supply chain constructs to respond to contemporary aspirations for society. This chapter also discusses the issues and challenges involved in adapting project-based supply chains for sustainability legislative oversight, and the development of sustainable supply chain practices enabled by digital technology, to propose alternative practices for project-based industries overall.

Chapter "Cold Chain and Its Application", the first chapter of the second part, describes cold chain logistics. The primary focus is given to the temperature-sensitive pharmaceutical products, vaccines, distribution practices, and quality management aspects for the cold chain technology. The cold chain technology seems to be an essential aspect of preserving food products, perishables, and delivering them in the right conditions in the open market. Any changes in the supply chain regarding time, temperature, distance, and environmental conditions result in the decline in the net present value of the products. This chapter deals with all the factors responsible for the quality assurance of the cold chain products giving a complete idea about the mechanism, processes, and monitoring functions, operational conditions for the same. This chapter provides special insights on the temperature standards, and other regulations for the cold chains. The chapter concludes with the future aspects related to this technology to produce, store, and handling all the produced goods.

Chapter "To Analyse the Impact and Benefits of Cold Chain Applications for Frozen Food at High-Temperature Zone: A Case Study of Rajasthan, India" provides an application of cold chain in a very hot temperature zone. There is a requirement not only for the cold chain system but also there is a growing need for modern incubation centers, quality testing laboratories, and storage amenities for enhancing the process of manufacturing. To cater to the needs of the people, the cold chain system for processed food became highly recommendable. In this chapter, Rajasthan, a North-Western province of India was selected for the study as it is known for its varied climate. The chapter observes the effectiveness of cold chain on edibles to retain its quality and to be available for the ultimate customer.

In chapter "Block Chain and Its Application", the only chapter of the third part, the overall conception of Block Chain technology, its recent applications have been targeted. It has also been tried to explore some other implementations of the related protocols. At the initial stage, discussions on the Bitcoin (the first-ever application of the technology) are being provided. Block Chain technology is known as the technological basis on which Bitcoin is built. This technology has made it possible that every kind of transaction is executed in a proper manner of decentralized way. It does not require any trusted kind of the third party in between the whole transaction. By offering distributed, undisputable, public verified records of the transactions, the Block Chain technology potentials to renovate many industries. In this chapter, a critical and detailed survey has been performed aiming towards the area of application of the Block Chain technology, its impact on the society, and its possible use-cases. The ideas purpose of this chapter is to provide that much ability to the readers to familiarize, to understand the state of art of the Block Chain world, in both the society and technology also.

Chapter "Inventory Modeling and Inventory Control Application", the commencing chapter of the last part of the book, i.e., the fourth part, concentrates on inventory modeling. It provides a brief description of the developmental history of dynamic inventory models and reviews different inventory control strategies, illustrating modern inventory control method via dynamical system approach. The chapter concludes with application areas and usefulness of the models reported from real-life inventory control point of view.

The concluding chapter of the part and the book, chapter "Radial Data Envelopment Analysis Approach to Performance Measurement: Study on Indian Banking System", concentrates on Data Envelopment Analysis. Altering customer demands with augmented competitive pressures significantly change the operating environment of banks in general and Indian banks in particular. Being competitive, Indian banks are continuously keeping themselves on a platform of performance as set by them and traditional methods for evaluation of performance are not working properly due to different issues as floated in this sector. This study adopts a non-traditional method of measuring performance via Data Envelopment Analysis; a frontier-based radial approach to measure performance between public and private sector Indian banks for the last half-decade. Using nonparametric test, this study also tests the hypothesis that "*there exist no statistical efficiency differences between the public and private sector banks*". Analyses show a significant result that: *no bank is found to be overall technically efficient with all banks scoring less than one, and the private sector banks is 1.5 percent more technically efficient than public sector banks*. Most significantly, strong ownership effect is proofed to be significant on the performance of Indian banking system.

First and foremost, the editors would like to thank God. It was with his blessings that this work could be completed to their satisfaction. You have given the power to believe in passion, hard work, and pursue dreams. The Editors could never have done this herculean task without the faith they have in you, the Almighty. They are thankful for this.

The editors would also like to thank all the chapter Contributors, the Reviewers, the Editorial Advisory Board Members, Book Development Editor, and the team of Publisher Springer Nature for their availability for work on this editorial project.

Throughout the process of editing this book, many individuals, from different walks of life, have taken time out to help. Last, but definitely not least, the editors would like to thank them all, their well-wishers, for providing them encouragement. They would have probably given up without their support.

Ranchi, India Kaushik Kumar
Aveiro, Portugal J. Paulo Davim

Contents

Application

About the Editors

Kaushik Kumar B.Tech. (Mechanical Engineering, REC (Now NIT), Warangal), MBA (Marketing, IGNOU) and Ph.D. (Engineering, Jadavpur University), is presently an Associate Professor in the Department of Mechanical Engineering, Birla Institute of Technology, Mesra, Ranchi, India. He has 14 years of teaching and research and over 11 years of industrial experience in a manufacturing unit of global repute. His areas of teaching and research interest are Quality Management Systems, Optimization, Non-conventional machining, CAD/CAM, Rapid Prototyping and Composites. He has 9 Patents, 28 books, 17 Edited Book Volume, 48 Book Chapters, 146 international Journal, 21 International and 1 National Conference publications to his credit. He is on the editorial board and review panel of seven International and one National Journals of repute. He has been felicitated with many awards and honours.

J. Paulo Davim received his Ph.D. in Mechanical Engineering in 1997, M.Sc. in Mechanical Engineering (materials and manufacturing processes) in 1991, Mechanical Engineering degree (5 years) in 1986 from the University of Porto (FEUP), the Aggregate title (Full Habilitation) from the University of Coimbra in 2005, and the D.Sc. (Higher Doctorate) from London Metropolitan University in 2013. He is a Senior Chartered Engineer by the Portuguese Institution of Engineers with an MBA and Specialist titles in Engineering and Industrial Management as well as in Metrology. He is also Eur Ing by FEANI-Brussels and Fellow (FIET) of IET-London. Currently, he is Professor at the Department of Mechanical Engineering of the University of Aveiro, Portugal. He is also distinguished as an honorary professor in several universities/colleges. He has more than 30 years of teaching and research experience in Manufacturing, Materials, Mechanical, and Industrial Engineering, with special emphasis in Machining and Tribology. He has also interest in Management, Engineering Education, and Higher Education for Sustainability. He has guided large numbers of Postdoc, Ph.D., and Master's students as well as coordinated and participated in several financed research projects. He has received several scientific awards and honors. He has worked as an evaluator of projects for ERC-European Research Council and other international research agencies as well as examiner of Ph.D. thesis for many universities in different countries. He is the Editor in Chief of several international journals, Guest Editor of journals, books Editor, book Series Editor, and Scientific Advisory for many international journals and conferences. Presently, he is an Editorial Board member of 30 international journals and acts as a reviewer for more than 100 prestigious Web of Science journals. In addition, he has also published, as editor (and co-editor), more than 150 books and as an author (and co-author) more than 15 books, 100 book chapters, and 500 articles in journals and conferences (more than 250 articles in journals indexed in Web of Science Core Collection/h-index 55+/9500+ citations, SCOPUS/h-index 60+/12000+ citations, Google Scholar/h-index 77+/19500+ citations).

Supply Chain

An Overview of Supply Chain Dynamics from a Behavioral Operations Perspective

Carmela Di Mauro, Salvatore Cannella, Roberto Dominguez, and Alessando Ancarani

Abstract This chapter provides an overview of the state of the art in Behavioral Operations Management studies focusing on supply chain dynamics. Specifically, we show how, in the last decade, this field of study has been gaining momentum in scholarly research. Additionally, we focus on supply chain dynamics and perform a systematic literature review on the behavioral causes of the bullwhip effect. Finally, we offer an illustration of how multi-method research can enrich behavioral operations management by presenting the results of a study in which a behavioral cause of supply chain dynamics, namely risk aversion is explored by combining human experiments and simulation.

Keywords Behavioral Operations Management · Supply Chain · Bullwhip Effect · Systematic Literature Review · Human Experiments

1 Behavioral Operations Management

Operations Management (OM henceforth) has traditionally been built around the concept that humans taking decisions concerning operations are fully informed and rational decision-makers. OM models typically incorporate the assumption that decision outcomes are independent of who the decision-maker is since the decision-maker

C. Di Mauro · S. Cannella (✉) · A. Ancarani
Management Engineering Research Group, DICAR—Department of Civil Engineering and Architecture, University of Catania, Via A. Doria 5, 95131 Catania, Italy
e-mail: salvatore.cannella@unict.it

C. Di Mauro
e-mail: cdimauro@unict.it

A. Ancarani
e-mail: alessandro.ancarani@unict.it

S. Cannella · R. Dominguez
Industrial Management Research Group, Industrial, Management and Business Administration Department, School of Engineering, University of Seville, Ave. Descubrimientos s/n, E41092 Seville, Spain
e-mail: rdc@us.es

K. Kumar and J. P Davim (eds.), *Supply Chain Intelligence*, Management and Industrial Engineering, https://doi.org/10.1007/978-3-030-46425-7_1

is deterministic and predictable in actions, self-interested, and emotionless (Boudreau et al. 2003).

The archetype behind this view is that of *Homo Oeconomicus*, a fully optimizing decision-maker with unlimited ability at problem-solving. However, this model has been questioned and overturned at least since Prospect Theory made its appearance (Kahneman and Tversky 1979). Working at the interface between economics and experimental psychology, Kahneman and Tversky showed that decisions generally engender heuristics and biases. A bias is an observed systematic deviation in decision-making from rationality, while a heuristic is a rule of thumb used to make decisions. Thus, a bias primarily generates deviations in outcomes of decisions from optimality, while a heuristic describes deviations in decision-making processes. The literature offers many illustrations of the systematic nature of some of the best-known decision biases: for instance, monetary gains and losses are evaluated with respect to a subjective reference point; equivalent gains and losses do not cancel out in the mind of the decision-maker because individuals are averse to losses; probabilities are always distorted when making decisions under risk. The existence of biases in human cognition and the application of heuristics have implications on decision models that are as far-reaching as Simon's (1957) concept of bounded rationality. In fact, they imply that assumptions underlying human behavior need re-thinking and that there is a need for training decision-makers to de-bias their standard way of thinking in order to improve the effectiveness and efficiency of their decisions.

In the last 30 years, cross-fertilization among economics, management, psychology, and sociology have led to the discovery that systematic violations of the standard paradigm of decision-making are rooted not only in cognitive failures but also concern other basic tenets of behavior. For instance, self-interested and opportunistic behavior, which forms the backbone of well-known theories such as Transaction Cost Economics (Williamson 1989), is contradicted by empirical findings documenting cooperative behavior and reciprocity (Fehr and Gachter 2000; Hoffman et al. 1998).

Although OM scholars have been slow in embracing the behavioral paradigm, the stream within OM denominated Behavioral Operations Management (BOM henceforth) has been constantly growing in the last 20 years. BOM calls for analyses of operations that have sound behavioral bases and that assign importance to factors such as the decision-maker's risk preferences (Bendoly et al. 2006), human cognitive processes and systematic biases (Gino and Pisano 2008; Narayanan and Moritz 2015), social preference stances (Loch and Wu 2008), and social norms.

In the editorial to an early special issue of the Journal of Operations Management dedicated to BOM, Bendoly et al. (2006) offered a framework for identifying OM problems worthy of behavioral investigation. These problems are classified using the "Intentions-Actions-Reactions" framework. "Intentions" encompass hypotheses about the decision-maker's objective function, monetary and non-monetary goals, and risk preferences. For instance, risk-seeking or risk-averse preferences are expected to impact decisions and outcomes (Eeckhoudt et al. 1995). As another example, the objective function of the decision-maker may incorporate non-monetary arguments such as trust, reciprocity, and fairness. "Actions" refer to human behavior in the model and encompass decision, communication, and motivation. "Reactions" capture human response to environmental modifications and include not only responses to feedbacks, and to game changes, but also learning processes and repeated social interactions.

Because many operations take place in complex and dynamic settings, behavioral analyses of OM need to address not only the cognitive foundations of individual decisions but need to encompass communication processes, coordination models, network formation, and social norms (Camerer and Malmendier 2007), as well as their impact on the team and organizational decisions.

The BOM literature has been gaining momentum since the early years two thousand. Between 2000 and 2010, several studies were published in major academic journals providing the foundations of BOM and suggesting future research directions (Bendoly et al. 2006, 2017; Gino and Pisano 2008; Loch and Wu 2005 among others). Several key journals in OM have devoted special issues to BOM (Bendoly et al. 2006; Gans and Croson 2008; Eckerd and Bendoly 2011; Croson et al. 2013; Ancarani et al. 2013). Other contributions have pinpointed appropriate research methodologies for the investigation of behavioral issues, such as laboratory human experiments (Croson and Donohue 2002), scenario-based experiments (Rungtusanatham et al. 2011), and system dynamics (Größler et al. 2008).

More recently, structured literature reviews showcasing the state of the art on specific research themes within BOM have been published. Kundu et al. (2015) used semantic latent analysis to identify research on behavioral supply chains. Greasley and Owen (2018) undertook a systematic literature review on modeling of human behavior in OM in discrete-event simulations. Fahimnia et al. (2019) published a structured literature review of BOM research published in eight key operations and supply chain management journals. To the present date, popular BOM research areas include inventory management, supply chain management, and procurement, although other areas such as service operations, revenue management, project management, and quality management are gaining momentum.

With the growth of the body of knowledge in BOM, the need for reviews addressing more focused areas of research within the discipline is felt. In this chapter, we present an overview of BOM literature on multilayered supply chain dynamics. The focus will be on behavioral aspects of supply chain instability and of the so-called "bullwhip effect", a favorite theme in system dynamics and one of the earliest topics to be addressed using the BOM lens.

The chapter is structured as follows. Section 2 briefly overviews the concept of supply chain instability and bullwhip effect, highlighting causes, consequences, and solutions. Section 3 presents a structured literature review of behavioral analysis of the bullwhip effect and supply chain dynamics. Section 4 summarizes the result of a multi-method behavioral study undertaken by the authors. Finally, Sect. 5 concludes and suggests future research directions.

2 Supply Chain Dynamics and Bullwhip Effect

Supply chain dynamics (Towill 1991; Goltsos et al. 2018) is a field of OM devoted to the study of time-varying phenomena in supply chains, such as the demand amplification phenomenon, also known in the literature as the "bullwhip effect" (Lee et al. 1997). This effect produces a detrimental increment in the variability of orders placed by the echelons of a supply chain, as order information is passed upstream. The negative economic consequences have been estimated to total as much as 30% of factory gate profits (Turrisi et al. 2013). In fact, negative consequences include holding cost increase (Ponte et al. 2018), lost sales and low service levels, limited productivity (Disney and Lambrecht 2008), increment in the amount of reschedules (Fioriolli and Fogliatto 2008), increased investment in capacity (Cannella et al. 2018) and unproductive use of transport capacity (Potter and Lalwani 2008). As pointed out by Bray and Mendelson (2012) in a study of a sample of 4689 public U.S. companies over the years 1974–2008, the scale of the problem is such that about two-thirds of firms are affected by the bullwhip effect. These figures have been confirmed by Shan et al. (2014) for another top world economy: based on financial data from over 1200 Chinese firms for the period 2002–2009, these authors found that instability of orders is larger than the volatility of sales in 67% of the firms analyzed. Furthermore, the recent trade downfall has generated an increment of the demand amplification in numerous manufacturing areas (Cannella et al. 2014). As showed by the European Central Bank and by the European Bank for Reconstruction and Development Working, such volatility has severely compromised the performance of several enterprises both in developed and emergent countries (Altomonte et al. 2012; Zavacka 2012). Thus, the bullwhip can be reasonably considered the "Sword of Damocles" of supply chains (Cannella et al. 2013).

Forrester (1961) first analyzed the bullwhip effect adopting the "industrial dynamics" approach. Since then, several studies have been undertaken to understand this perilous effect. The bullwhip has been typically attributed to operational causes such as lead time, rationing and shortage gaming order batching, demand forecasting, price fluctuation (Lee et al. 1997), capacity limits, number of echelons, local optimization without global vision, poor synchronization, improper control system, multiplier effect, misperception of feedback, replenishment policy, inventory policy, and lack of transparency (Bhattacharya and Bandyopadhyay 2011).

However, since Sterman (1989) behavioral causes have also been pinpointed to understand the bullwhip effect. Sterman, using the "beer distribution" or "beer" business game, attributed the amplification of order variance to the tendency of members of the supply chain to overlook the inventory-on-order (i.e., the orders placed but not yet received), a cause of amplification signaling "irrational behavior" (Wang and Disney 2016). Sterman's seminal study led to revise the analysis of the demand amplification and to acknowledge that its causes are twofold—operational causes and behavioral causes (Bhattacharya and Bandyopadhyay 2011). According to Bhattacharya and Bandyopadhyay (2011), behavioral causes include Fear of empty stock, lack of learning and/or training, and neglecting time delays in making ordering decisions. However, other behavioral causes have been unveiled by later studies.

In the following section, we present a systematic literature review on behavioral studies of the bullwhip effect. Studies will be classified by research methodology, supply chain structure, and behavioral issues investigated.

3 Literature Review

3.1 Search Process

A systematic review of the literature was carried out using the framework laid out by Denyer and Tranfield (2009). This well-known methodology relies on some key steps in order to reduce the researcher's bias in identifying and analyzing scholarly documents and facilitate structured, unambiguous, and clear communication of results. The following steps were followed in this study: (1) question formulation, (2) source selection, (3) source evaluation, and (4) data analysis.

The first step "Question formulation" defines a clear research question. In this work, the aim is to analyze the behavioral factors that influence the bullwhip effect. Through this review, it will be possible to build a summary of major behavioral triggers of the bullwhip effect and highlight future research directions. Therefore, we pursue the following research question:

RQ. What aspects of supply chain dynamics and the bullwhip effect have been explored using a behavioral lens?

Concerning the "source selection step", we considered contributions published in peer-reviewed academic journals, international books, and conference proceedings in the English language. In order to include potential contributions from various disciplinary areas, a keyword search without a reduction in the scope of the disciplinary area of journals was adopted. The search coverage period included documents published up to the end of 2018.

Next, topic-related keywords were used to search for documents in the Scopus and Web of Science databases. The following two search strings were adopted in title, keywords and abstract:

- ("bullwhip effect" OR "Forrester" OR "order amplification" OR "demand amplification" OR "supply chain dynamics") AND ("behavioral" OR "behavioral" OR "human behavior" OR "human behavior");
- ("bullwhip effect" OR "Forrester" OR "order amplification" OR "demand amplification" OR "supply chain dynamics") AND ("beer game" OR "beer distribution game").

This search process led to the identification of 40 documents that were read in full.

In order to evaluate the relevance of the documents found, we proceeded with the definition of the exclusion criteria. Articles were excluded from subsequent analysis whenever one of the following issues applied: (i) lack of insights on specific behavioral issues; (ii) research replicated previous findings, therefore not providing any new contribution; (iii) the article was a "Pink Elephant", i.e. presented the search string words only in the keywords or abstract; (iv) an extended version of the same research has been published later in a higher quality international journal. This exclusion process produced a final list of 21 articles.

3.2 Findings

Contributions included in the analysis (Table 1) were classified by:

(i) year of publication;
(ii) research methodology adopted in the study;
(iii) supply chain structures;
(iv) behavioral issue addressed;
(v) performance metrics used.

Here following, we briefly discuss results with respect to supply chain structures envisaged, behavioral issues addressed, and methodologies used.

Supply chain structures The preferred framework for the analysis of behavioral issues affecting the bullwhip effect is the so-called "beer game" framework (Forrester 1961; Sterman 1989), a classic business game in which the supply chain consists of four echelons (retailer–wholesaler–distributor–factory). During the game, orders are placed by each participant to the immediate upstream supplier at each period, and orders of the downstream customer are filled. Typically, an information lead time exists at each supply chain echelon when a buyer places an order before this latter is known to the upstream supplier. In addition, a transportation lead time is included for the shipment of goods ordered to the downstream echelon. The same parameters of the game (i.e. lead times) apply al all echelons, including the factory when beer is produced. Goods received at time t correspond to those shipped by the upstream supplier earlier on. The supplier can partially fulfill an order, depending on the supplier's inventory availability. Inventory holding costs and backlog costs for failing to fulfil orders apply at every echelon (Sterman 1989).

Most of the extant behavioral literature using the Beer Game framework has analyzed orders placed in the context of demand uncertainty. For instance, the effect of varying demand distribution has been explored (Croson and Donohue 2006; Wu and Katok 2006). Demand uncertainty can be regarded as the baseline type of beer game, on the ground that uncertainty stemming from the demand side is difficult to diversify away for firms. However, firms undoubtedly have to cope with supply-side uncertainty (Tang and Tomlin 2008). In past research of our research team,

Table 1 Literature review results

	Methodology	Supply chain structure (echelons)	Behavioral issue	Performance metrics
Pamulety and Pillai (2011)	Human experiments	K = 4	• Impact of imperfect demand information	• Variance of orders • Fill rate • Total inventory • Total holding costs
Croson and Donohue (2005)	Human experiments	K = 4	• Impact of access to upstream and downstream inventory information	• Average variance of orders
Haines et al. (2017)	Human experiments	K = 4	• Impact of additional information	• Total costs of the supply chain
Thompson and Badizadegan (2015)	Simulation	K = 4	• Impact of available information versus perfect information	• Total costs of the supply chain
Croson and Donohue (2006)	Human experiments	K = 4	• Impact of use inventory information	• Variance of orders
Banbury et al. (2010)	Simulation	K = 4	• Analysis of the role of collaboration	• Total cost • Orders • Back orders • Bullwhip
De Almeida et al. (2015)	Systematic literature review	N.A.	• Impact of trust and collaboration	N.A.
Wu and Katok (2006)	Human experiments	K = 4	• Analysis of experiential learning, systems learning, and organizational learning	• Standard deviation of orders
Delhoum and Scholz-Reiter (2009)	Human experiments	K = 2	• Learning inventory control	• Marginal costs
Zhao and Zhao (2015)	Human experiments	K = 5	• Analysis of the behavioral biases into different information scenarios	• Standard deviation of orders • Average operating cost
Udenio et al. (2017)	Control theory	K = 1	• Smoothing/over-reaction to inventory and pipeline mismatches • Under/overestimation of pipeline	• Order and inventory variations
Narayanan and Moritz (2015)	Human experiments	K = 4	• Impact of individual cognitive reflection	• Order variance • Average supply chain costs
Li and Yan (2015)	Simulation	K = 2	• Impact of lack of trust, incentive misalignment and risk aversion into decision-making process	• Order variance • Service level
Ancarani et al. (2013)	Human experiments	K = 4	• Reaction to demand versus supply uncertainty and uncertainty attitude	• Orders • Inventory holding costs • Backlogging costs • Total costs

(continued)

Table 1 (continued)

	Methodology	Supply chain structure (echelons)	Behavioral issue	Performance metrics
Croson et al. (2014)	Human experiments	K = 4	• Coordination risk	• Standard deviation of orders • Inventory
Hung and Ryu (2008)	Human experiments	K = 4	• Impact of different risk appetite of inventory managers with respect to demand changes and supplier failures	• Order deviation
Nienhaus et al. (2006)	Human experiments	K = 4	• Impact of safe harbor and panic	• Lead time • Capital costs of a supply chain
Croson and Donohue (2003)	Simulation	K = 4	• Impact of point of sale (POS) data sharing on ordering decisions	• Order oscillations
Bruccoleri et al. (2014)	Simulation	K = 1	• Impact of worker's stress on inventory record inaccuracy	• F* range: proxy of worker psychological stability to stress and pressure • F*: the midpoint of throughput optimal range
Wang et al. (2014)	Numerical experiments	K = 2	• Impact of consumer price forecasting behavior	• Market demand price • Price sensitivity • Standard deviation of the price
Niranjan et al. (2009)	Human experiments	K = 4	• Review of Sterman's paper and analysis of the impact of inventory and penalty costs	• Order variance ratio • Cost per period

we have used supply-side uncertainty scenarios within the beer game and made it operational through stochastic lead times (Ancarani et al. 2013, 2016). In Ancarani et al. (2013), comparison of a treatment with supply-side uncertainty with another in which there is both demand and supply-side uncertainty shows that orders and inventory decrease when the overall uncertainty of the system increases. This finding runs counter intuition and numerical simulations that would predict an increase in orders and is explained using the gain-loss framework of Tversky and Kahneman (1992).

Main behavioral issues addressed While several works have been devoted to the study of remedies to the operational causes of the bullwhip effect (e.g., information sharing and collaboration in supply chain, smoothing replenishment rules, improved forecasting, etc.) (Disney and Lambrecht 2008), it seems that behavioral causes of the bullwhip effect and how they can be analyzed and avoided have received less attention. The literature review has identified the following behavioral causes of supply chain dynamics and the bullwhip effect: coordination risk (Croson et al. 2014), stress of workers (Bruccoleri et al. 2014), individual cognitive reflection in decision-making process (Narayanan and Moritz 2015), risk preferences (Hung and

Ryu 2008), uncertainty aversion and preference (Ancarani et al. 2013), and lack of trust (De Almeida et al. 2015; Li and Yan 2015).

Overall, the articles selected suggest that there are three broad themes on which the BOM literature has focused: the role of information sharing inside the supply chain (Croson and Donohue 2003, 2005), the importance of learning (Wu and Katok 2006), and the risk-taking behavior stemming from individual cognitive traits (Ancarani et al. 2013; Hung and Ryu 2008).

Research methodologies Analysis of the research methodologies used shows that human experiments in controlled settings have become the elective approach for investigating behavioral issues in the supply chain (13 documents). Using human experiments for the collection of primary data in management and operations has now been accepted as an appropriate methodology (Bendoly et al. 2006; Boyer and Swink 2008). The main advantage of human experiments is the high degree of internal validity. On the other hand, external validity often runs low, because the use of students as experiment participants is generally accepted. In addition, human experiments provide a valued tool for observing human decisions in an incentive-compatible environment, adding reliability and validity to these measures with respect to standard survey methodologies. There are several respects in which human experiments can inform OM as to the relevance of behavioral factors (Croson and Donohue 2002): (i) by studying specific behavioral factors while keeping operational variables constant; (ii) by studying operational and organizational factors, assuming invariance of behavioral characteristics; (iii) by studying the interactions between operational and behavioral factors.

To date, there is still a scarcity of studies combining multiple methodologies to investigate behavioral issue and, in particular, combining empirical methodologies such as human experiments with numerical experiments and "what if" simulations. The next section illustrates the advantages of a multi-method approach to BOM by presenting the results of an earlier study by our research team.

4 Multi-method Studies in BOM

Herein, we present the results of a recent multi-method study that addresses a potential behavioral cause of the bullwhip effect, i.e., the risk aversion of the decision-maker. Please refer to Cannella et al. (2019) for a comprehensive account of the study.

The study belongs to the group of research contributions studying specific behavioral factors while keeping operational variables constant. In the case at hand, the behavioral issue is represented by risk aversion. Risk-averse individuals wish to avoid failures and this desire overrides that to achieve good outcomes (Lopes 1987). Risk aversion may be considered akin to a personality trait that leads the decision makes to be willing to accept a lower but sure outcome, instead of a higher but risky outcome.

Using a four-echelon Beer game, the study explored which factors of the replenishment rule can be affected by risk aversion and how the individual risk aversion may impact on inventory holding costs, and how the location of risk-averse members within a multi-echelon supply chain impact the supply chain performance. The analysis was carried out by adopting a multi-method study (see e.g., Adner and Levinthal 2001; Evers and Wan 2012; Van Oorschot et al. 2013; Chandrasekaran et al. 2016), by which human experiments and multi-agent simulation were combed in a two-step procedure. Human experiments were adopted as they permit the generation of primary data within a highly controlled environment (the lab), thus ensuring that causal inferences can be made (Croson and Donohue 2002). Moreover, human experimentation permits disaggregating participants' decisions according to the behavioral or cognitive characteristic of interest. Prior to observing the decision-makers inventory decisions, this technique allows the use of a controlled setting in order to measure their risk aversion. However, a great amount of experimental data points would be required to perform a complete investigation of the effect of risk aversion on the performance of a supply chain using human experiments (e.g., to systematically study the impact of several degrees of risk aversion of its members on supply chain performance). Thus, given that a complete investigation of risk aversion on a supply chain would be challenging in terms of data requirements, the human experiment was complemented with computer simulation tests (Chandrasekaran et al. 2016). Specifically, we combined the human experiment with multi-agent simulation, a common methodology in supply chain studies and providing a means to address complexity (Wu et al. 2015). Multi-agent modeling is considered a powerful tool for the analysis of the dynamics of supply chains, offering an excellent instrument for modeling the independence, communication, organization, and decision processes of supply chains (see e.g., Chatfield et al. 2012; Dev et al. 2016; Yu et al. 2017; Ponte et al. 2017, 2018; Dominguez et al. 2018; etc.).

Here following, we provide an account of how the two methods were combined in the study in order to improve the understanding of behavior and provide managerial recommendations for supply chain management. In the first stage, in accordance with other supply chain experiments (Croson et al. 2014), we obtained data on inventory decisions in a four-echelon chain by adopting laboratory human experiment in which supply chain professionals participated as human subjects. Individual risk aversion of the participants was measured prior to the beginning of the experiment and outcomes from the human experiment were used to understand how variations in individual risk aversion impact the behavior and performance of the different supply chain members. Results from the human experiment suggested a working hypothesis, namely that higher risk aversion generates larger inventory holdings due to the adoption of higher safety stock.

In the second stage of the study, the working hypothesis generated through the human experiment outcomes were used to build a multi-agent simulation model, which explored the behavior of supply chain performance, simulating different degrees of risk aversion for each member of the supply chain. Specifically, we simulated different combinations of risk aversion across the supply chain, modeled by assuming different safety stock factor settings. Impacts on the customer service level and operational performance of the simulated supply chain were observed. The operational performance was measured in terms of order rate variance ratio, namely the bullwhip effect (Disney and Lambrecht 2008) and inventory levels along the supply chain (Chatfield et al. 2004), while the customer service level was measure by assessing the fill rate values at the retailer.

By so doing, we adopted human experiments as a hypothesis generation tool and used multi-agent simulation to explore in greater depth the impact on the dynamics of a four-echelon chain. The relevance of behavioral issues in supply chain management was unveiled thanks to the multi-method approach adopted in this study, which added multidimensional insights to a growing line of investigation (Bendoly et al. 2010; Chandrasekaran et al. 2016). Specifically, results suggested that: (1) the bullwhip effect and inventory holding costs linearly worsen as the number of echelons run by highly risk-averse inventory managers increases. (2) Supply chains characterized by an equal number of low (or highly) risk-averse members show analogous bullwhip and inventory holding costs values, regardless of which echelons low (or high) risk aversion managers are assigned to. (3) A high customer service level can be achieved when there are risk-averse retailers, regardless of the risk aversion of the other members of the chain.

The above results could not have been achieved without using a multi-method approach. In fact, while the human experiment uncovered the link between risk aversion and inventory holdings, and suggested the link between risk aversion and the safety stock factor, the multi-agent simulation model threw light on the impact on the dynamics of the multi-echelon supply chain of positioning actors with different risk aversion levels at different echelons.

5 Final Remarks and Research Directions

In this chapter, we have provided an overview of the state of the art in BOM studies focusing on supply chain dynamics. Specifically, we traced the historical path of BOM and how, since the early years two thousand, this field of study of OM has been gaining the attention of researchers. Next, we introduced the concept of supply chain dynamics and the bullwhip effect, pointing the causes, consequences, and remedies to this detrimental phenomenon, with particular attention to its behavioral aspects. In addition, we presented a systematic literature review on the behavioral causes of the bullwhip effect by adopting the framework laid out by Denyer and Tranfield (2009). By doing so, we showed how BOM in supply chain dynamics is an emerging field of study, which deserves further research. Finally, we presented the results of a multi-method study in which we have explored the impact of a potential behavioral cause of the bullwhip effect, i.e., risk aversion. This study has allowed stressing the usefulness of adopting multiple methodological leans to address behavioral factors within OM and the supply chain.

Regarding the future directions of BOM in supply chain dynamics, we agree with the emerging topics identified by Fahimnia et al. (2019). In the next future, in order to understand supply chain dynamics more research is needed to address the behavior of supply chain members when disruptions occur. Analogously, behavioral buyer–supplier relations in sustainable, and closed-loop supply chain represent a further challenge for the BOM community. Finally, the impact of human judgement in partially automated supply chains and the interaction between ERP systems and management decisions deserves research attention.

Acknowledgements Alessandro Ancarani, Carmela Di Mauro, Roberto Dominguez, and Salvatore Cannella acknowledge financial support by the University of Catania within the project "Piano della Ricerca Dipartimentale 2016–2018" of the Department of Civil Engineering and Architecture. Furthermore, Roberto Dominguez and Salvatore Cannella appreciate the financial support of the Spanish Ministry of Science and Innovation, through the project PROMISE (reference DPI201680750P), and the University of Seville, through the V/VI PPIT-US.

References

Adner, R., & Levinthal, D. (2001). Demand heterogeneity and technology evolution: Implications for product and process innovation. *Management Science, 47*(5), 611–628.

Altomonte, C., Mauro, F. D., Ottaviano, G., Rungi, A., & Vicard, V. (2012). *Global value chains during the great trade collapse: A bullwhip effect?* (Working Paper Series 1412). European Central Bank.

Ancarani, A., Di Mauro, C., & D'Urso, D. (2013). A human experiment on inventory decisions under supply uncertainty. *International Journal of Production Economics, 142,* 61–73.

Ancarani, A., Di Mauro, C., & D'Urso, D. (2016). Measuring overconfidence in inventory management decisions. *Journal of Purchasing and Supply Management, 22*(3), 171–180.

Banbury, S., Helman, S., Spearpoint, J., & Tremblay, S. (2010). Cracking the bullwhip: Team collaboration and performance within a simulated supply chain. *Proceedings of the Human Factors and Ergonomics Society Annual Meeting, 54*(19), 1620–1624.

Bendoly, E., Donohue, E., & Schultz, K. L. (2006). Behaviour in operations management: Assessing recent findings and revisiting old assumptions. *Journal of Operations Management, 24*(6), 737–752.

Bendoly, E., Croson, R., Goncalves, P., & Schultz, K. (2010). Bodies of knowledge for research in behavioral operations. *Production and Operations Management, 19*(4), 434–452.

Bendoly, E., McClintock, A., & Pandey, R. (2017). Human behavior in operations. In *The Routledge companion to production and operations management* (pp. 326–342). Routledge.

Bhattacharya, R., & Bandyopadhyay, S. (2011). A review of the causes of bullwhip effect in a supply chain. *The International Journal of Advanced Manufacturing Technology, 54*(9–12), 1245–1261.

Boudreau, J., Hopp, W., McClain, J. O., & Thomas, L. J. (2003). On the interface between operations and human resources management. *Manufacturing and Service Operations Management, 5*(3), 179–202.

Boyer, K. K., & Swink, M. L. (2008). Empirical elephants—why multiple methods are essential to quality research in operations and supply chain management. *Journal of Operations Management, 26*(3), 338–344.

Bray, R. L., & Mendelson, H. (2012). Information transmission and the bullwhip effect: An empirical investigation. *Management Science, 58*(5), 860–875.

Bruccoleri, M., Cannella, S., & La Porta, G. (2014). Inventory record inaccuracy in supply chains: The role of workers' behavior. *International Journal of Physical Distribution & Logistics Management, 44*(10), 796–819.

Camerer, C., & Malmendier, U. (2007). Behavioral economics of organizations. *Behavioral Economics and its Applications*, 235-290.

Cannella, S., Barbosa-Póvoa, A. P., Framinan, J. M., & Relvas, S. (2013). Metrics for bullwhip effect analysis. *Journal of the Operational Research Society, 64*(1), 1–16.

Cannella, S., Ashayeri, J., Miranda, P. A., & Bruccoleri, M. (2014). Current economic downturn and supply chain: The significance of demand and inventory smoothing. *International Journal of Computer Integrated Manufacturing, 27*(3), 201–212.

Cannella, S., Dominguez, R., Ponte, B., & Framinan, J. M. (2018). Capacity restrictions and supply chain performance: Modelling and analysing load-dependent lead times. *International Journal of Production Economics, 204,* 264–277.

Cannella, S., Di Mauro, C., Dominguez, R., Ancarani, A., & Schupp, F. (2019). An exploratory study of risk aversion in supply chain dynamics via human experiment and agent-based simulation. *International Journal of Production Research, 57*(4), 985–999.

Chandrasekaran, A., Linderman, K., Sting, F. J., & Benner, M. J. (2016). Managing R&D project shifts in high-tech organizations: A multi-method study. *Production & Operations Management, 25*(3), 390–416.

Chatfield, D. C., Kim, J. G., Harrison, T. P., & Hayya, J. C. (2004). The bullwhip effect—Impact of stochastic lead time, information quality, and information sharing: A simulation study. *Production and Operations Management, 13*(4), 340–353.

Chatfield, D. C., Hayya, J. C., & Cook, D. P. (2012). Stockout propagation and amplification in supply chain inventory systems. *International Journal of Production Research, 51*(5), 1491–1507.

Croson, R., & Donohue, K. (2002). Experimental economics and supply-chain management. *Interfaces, 32*(5), 74–82.

Croson, R., & Donohue, K. (2003). Impact of POS data sharing on supply chain management: An experimental study. *Production and Operations Management, 12*(1), 1–11.

Croson, R., & Donohue, K. (2005). Upstream versus downstream information and its impact on the bullwhip effect. *System Dynamics Review: The Journal of the System Dynamics Society, 21*(3), 249–260.

Croson, R., & Donohue, K. (2006). Behavioral causes of the bullwhip effect and the observed value of inventory information. *Management Science, 52,* 323–336.

Croson, R., Schultz, K., Siemsen, E., & Yeo, M. L. (2013). Behavioral operations: The state of the field. *Journal of Operations Management, 31*(1–2), 1–5.

Croson, R., Donohue, K., Katok, E., & Sterman, J. (2014). Order stability in supply chains: Coordination risk and the role of coordination stock. *Production and Operations Management, 23*(2), 176–196.

De Almeida, M. M. K., Marins, F. A. S., Salgado, A. M. P., Santos, F. C. A., & da Silva, S. L. (2015). Mitigation of the bullwhip effect considering trust and collaboration in supply chain management: A literature review. *The International Journal of Advanced Manufacturing Technology, 77*(1–4), 495–513.

Delhoum, S., & Scholz-Reiter, B. (2009). The influence of decision patterns of inventory control on the bullwhip effect based on a simulation game of a production network. *Production Planning & Control, 20*(8), 666–677.

Denyer, D., & Tranfield, D. (2009). Producing a systematic review. In D. A. Buchanan & A. Bryman (Eds.), *The SAGE handbook of organizational research methods* (pp. 671–689). London: Sage.

Dev, N. K., Shankar, R., Gunasekaran, A., & Thakur, L. S. (2016). A hybrid adaptive decision system for supply chain reconfiguration. *International Journal of Production Research, 54*(23), 7100–7114.

Disney, S. M., & Lambrecht, M. R. (2008). On replenishment rules, forecasting, and the bullwhip effect in supply chains. *Foundations and Trends in Technology, Information and Operations Management, 2*(1), 1–80.

Dominguez, R., Cannella, S., Barbosa-Póvoa, A. P., & Framinan, J. M. (2018). Information sharing in supply chains with heterogeneous retailers. *Omega, 79,* 116–132.

Eckerd, S., & Bendoly, E. (2011). Introduction to the discussion forum on using experiments in supply chain management research. *Journal of Supply Chain Management, 47*(3), 3–4.
Eeckhoudt, L., Gollier, C., & Schlesinger, H. (1995). The risk-averse (and prudent) newsboy. *Management science, 41*(5), 786–794.
Evers, P. T., & Wan, X. (2012). Systems analysis using simulation. *Journal of Business Logistics, 33*(2), 80–89.
Fahimnia, B., Pournader, M., Siemsen, E., Bendoly, E., & Wang, C. (2019). Behavioral operations and supply chain management—A review and literature mapping. *Decision Sciences*. https://doi.org/10.1111/deci.12369.
Fehr, E., & Gachter, S. (2000). Fairness and retaliation: The economics of reciprocity. *The Journal of Economic Perspectives, 14*(3), 159–181.
Fioriolli, J. C., & Fogliatto, F. S. (2008). A model to quantify the bullwhip effect in systems with stochastic demand and lead time. In *IEEE International Conference on Industrial Engineering and Engineering Management* (pp. 1098–1102).
Forrester, J. W. (1961). *Industry dynamics*. Cambridge, MA.
Gans, N., & Croson, R. (2008). Introduction to the special issue on behavioral operations. *Manufacturing & Service Operations Management, 10*(4), 563–565.
Gino, F., & Pisano, G. (2008). Toward a theory of behavioral operations. *Manufacturing & Service Operations Management, 10*(4), 676–691.
Goltsos, T. E., Ponte, B., Wang, S., Liu, Y., Naim, M. M., & Syntetos, A. A. (2018). The boomerang returns? Accounting for the impact of uncertainties on the dynamics of remanufacturing systems. *International Journal of Production Research,* 1–34. https://doi.org/10.1080/00207543.2018.1510191.
Greasley, A., & Owen, C. (2018). Modelling people's behaviour using discrete-event simulation: A review. *International Journal of Operations & Production Management, 38*(5), 1228–1244.
Größler, A., Thun, J.-H., & Milling, P. M. (2008). System dynamics as a structural theory in operations management. *Production and Operations Management, 17*(3), 373–384.
Haines, R., Hough, J., & Haines, D. (2017). A metacognitive perspective on decision making in supply chains: Revisiting the behavioral causes of the bullwhip effect. *International Journal of Production Economics, 184,* 7–20.
Hoffman, E., McCabe, K. A., & Smith, V. L. (1998). Behavioral foundations of reciprocity: Experimental economics and evolutionary psychology. *Economic Inquiry, 36*(3), 335–352.
Hung, K. T., & Ryu, S. (2008). Changing risk preferences in supply chain inventory decisions. *Production Planning and Control, 19*(8), 770–780.
Kahneman, D., & Tversky, A. (1979). Prospect theory: An analysis of decision under risk. *Econometrica, 47*(2), 263–292.
Kundu, A., Jain, V., Kumar, S., & Chandra, C. (2015). A journey from normative to behavioral operations in supply chain management: A review using latent semantic analysis. *Expert Systems with Applications, 42*(2), 796–809.
Lee, H. L., Padmanabhan, V., & Whang, S. (1997). The bullwhip effect in supply chains. *Sloan Management Review, 38*(3), 93–102.
Li, Z., & Yan, G. (2015). Exploring different order decision behaviors with bullwhip effect and service level measures in supply chain system. *Discrete Dynamics in Nature and Society*.
Loch, C. H., & Wu, Y. (2005). Behavioral operations management. *Foundations and Trends® in Technology, Information and Operations Management, 1*(3), 121–232.
Loch, C., & Wu, Y. (2008). Social preferences and supply chain performance: An experimental study. *Management Science, 54*(11), 1835–1849.
Lopes, L. L. (1987). Between hope and fear: The psychology of risk. *Advances in Experimental Social Psychology, 20,* 255–295.
Narayanan, A., & Moritz, B. B. (2015). Decision making and cognition in multi-echelon supply chains: An experimental study. *Production and Operations Management, 24*(8), 1216–1234.

Nienhaus, J., Ziegenbein, A., & Schönsleben, P. (2006). How human behaviour amplifies the bullwhip effect. A study based on the beer distribution game online. *Production Planning & Control, 17*(6), 547–557.
Niranjan, T. T., Metri, B. A., & Aggarwal, V. (2009). The behavioural causes of the bullwhip effect: Breaking the mould. *International Journal of Services and Operations Management, 5*(3), 350.
Pamulety, T. C., & Pillai, V. M. (2011). Impact of information sharing in supply chain performance. In *Technology systems and management* (pp. 327–332). Berlin, Heidelberg: Springer.
Ponte, B., Sierra, E., de la Fuente, D., & Lozano, J. (2017). Exploring the interaction of inventory policies across the supply chain: An agent-based approach. *Computers & Operations Research, 78*, 335–348.
Ponte, B., Costas, J., Puche, J., Pino, R., & de la Fuente, D. (2018). The value of lead time reduction and stabilization: A comparison between traditional and collaborative supply chains. *Transportation Research Part E: Logistics and Transportation Review, 111*, 165–185.
Potter, A., & Lalwani, C. (2008). Investigating the impact of demand amplification on freight transport. *Transportation Research Part E: Logistics and Transportation Review, 44*(5), 835–846.
Rungtusanatham, M., Wallin, C., & Eckerd, S. (2011). The vignette in a scenario-based role-playing experiment. *Journal of Supply Chain Management, 47*(3), 9–16.
Shan, J., Yang, S., Yang, S., & Zhang, J. (2014). An empirical study of the bullwhip effect in China. *Production and Operations Management, 23*(4), 537–551.
Simon, H. A. (1957). Models of man; social and rational. Wiley
Sterman, J. D. (1989). Modeling managerial behavior: Misperceptions of feedback in a dynamic decision making experiment. *Management Science, 35*(3), 321–339.
Tang, C., & Tomlin, B. (2008). The power of flexibility for mitigating supply chain risks. *International Journal of Production Economics, 116*(1), 12–27.
Thompson, K. M., & Badizadegan, N. D. (2015). Valuing information in complex systems: An integrated analytical approach to achieve optimal performance in the beer distribution game. *IEEE Access, 3*, 2677–2686.
Towill, D. R. (1991). Supply chain dynamics. *International Journal of Computer Integrated Manufacturing, 4*(4), 197–208.
Turrisi, M., Bruccoleri, M., & Cannella, S. (2013). Impact of reverse logistics on supply chain performance. *International Journal of Physical Distribution & Logistics Management, 43*(7), 564–585.
Tversky, A., & Kahneman, D. (1992). Advances in prospect theory: Cumulative representation of uncertainty. *Journal of Risk and Uncertainty, 5*(4), 297–323.
Udenio, M., Vatamidou, E., Fransoo, J. C., & Dellaert, N. (2017). Behavioral causes of the bullwhip effect: An analysis using linear control theory. *IISE Transactions, 49*(10), 980–1000.
Van Oorschot, K. E., Akkermans, H., Sengupta, K., & Van Wassenhove, L. N. (2013). Anatomy of a decision trap in complex new product development projects. *Academy of Management Journal, 56*(1), 285–307.
Wang, X., & Disney, S. M. (2016). The bullwhip effect: Progress, trends and directions. *European Journal of Operational Research, 250*(3), 691–701.
Wang, N., Ma, Y., He, Z., Che, A., Huang, Y., & Xu, J. (2014). The impact of consumer price forecasting behaviour on the bullwhip effect. *International Journal of Production Research, 52*(22), 6642–6663.
Williamson, O. E. (1989). Transaction cost economics. *Handbook of industrial organization, 1*, 135–182.
Wu, D. Y., & Katok, E. (2006). Learning, communication, and the bullwhip effect. *Journal of Operations Management, 24*(6), 839–850.
Wu, D., Olson, D. L., & Dolgui, A. (2015). Decision making in enterprise risk management: A review and introduction to special issue. *Omega, 57*, 1–4.
Yu, Y., Cao, R. Q., & Schniederjans, D. (2017). Cloud computing and its impact on service level: a multi-agent simulation model. *International Journal of Production Research, 55*(15), 4341–4353.

Zavacka, V. (2012). *The bullwhip effect and the great trade collapse* (Working Paper No. 148). European Bank for Reconstruction and Development.
Zhao, Y., & Zhao, X. (2015). On human decision behavior in multi-echelon inventory management. *International Journal of Production Economics, 161*, 116–128.

Classification of Factors Associated with a Closed-Loop Supply Chain System, Their Modelling Methods and Strategies

Hridayjit Kalita, Kaushik Kumar, and J. Paulo Davim

Abstract Environmental impact of the industries has been one of the primary concerns of the society today due to increased carbon emission and non-sustainability in their open-end supply chain systems. Closed-loop supply chain system (CLSC) is an improved system often suggested by researchers to enhance sustainability and lower environmental degradation by closing the loop of the product lifecycle and encouraging product returns and other green activities. The concept was strengthened by the introduction of various policies and regulations by the government for the sake of maintaining the health of the environment. In this paper, a thorough investigation of the CLSC system and categories/groups of factors influencing the optimal parameters of a CLSC system have been reported. Modelling of CLSC strategies involves different uncertainty problems which are required to be solved using different mathematical models. These uncertainty problems are governed by the varying numbers assigned to the uncertainty factors. This paper gives an account of various modelling methods and solution strategies commonly used in the research of CLSC systems for solving the uncertainty problems and analyze different aspects of the practical system for an optimized supply chain system.

Keywords Closed-loop supply chain · Carbon emission · Remanufacturing · Uncertainty · Modelling · Stochastic method · Incentive mechanism

1 Introduction

With recent awareness regarding climate change and environmental degradation in industries, the focus on the topics such as sustainability and carbon emission criteria have become the key discussion for researchers and managers in their evaluation of green supply chain management and production system (Mohammed et al. 2017; Ma and Li 2018; Asim et al. 2019). Circular economy (CE) in a broader perspective

H. Kalita · K. Kumar (✉)
Birla Institute of Technology, Mesra, Ranchi, India
e-mail: kkumar@bitmesra.ac.in

J. P. Davim
University of Aveiro, Aveiro, Portugal

K. Kumar and J. P Davim (eds.), *Supply Chain Intelligence*, Management and Industrial Engineering, https://doi.org/10.1007/978-3-030-46425-7_2

for a closed-loop supply chain (CLSC) system has been found to be a hot research area which is evident from the work of a large number of researchers (Merli et al. 2018; Prieto-Sandoval et al. 2018). Circular economy (CE) as defined in Homrich et al. (2018) is a strategic system that opposes the traditional open-end system and incorporates sustainability into it by minimizing resource utilization, maximizing output profit, minimizing waste disposal and other environmental measures along with considering customer satisfaction as a major priority. The term CLSC was previously not introduced in the industries and only gained attention by researchers from public awareness (Dowlatshahi 2000) until governments later passed legislation that allowed the manufacturers to take measures on their end of life/use products before disposing off to enhance their productivity and at the same time reducing environmental impacts (Georgiadis and Besiou 2010; Neto et al. 2010). CLSC can be introduced from the literature as a system in design, operations and controls to enhance the value/profit creation throughout the entire lifecycle of a manufacturer's product and at the same time dynamically recovering the values over time (Guide and Van Wassenhove 2009).

Traditional open-end supply chain system kept the customer satisfaction as the major objective and involved entities in its framework such as manufacturer, suppliers, transporter, retailers, warehouses and the customers themselves (Govindan et al. 2020). This is generally called the forward supply chain. On the other hand, in reverse supply chains or reverse logistics (RL), the objective is to collect the end of life/used products from the consumers in order to reuse, recycle, repair, remanufacture for a cost-effective and efficient process and to dispose of properly if needed (Rogers and Tibben-Lembke 1999). When both forward and reverse supply chains are integrated together to create value throughout the entire lifecycle of the product the system changes to CLSC.

The characteristics of the CLSC needs to be categorized and classified to help researchers in creating new models (Souza 2013; Govindan et al. 2015) that can make the system a further efficient one. This classification helps the CLSC system in analyzing the behavioural trends in their rational players, choices of their stakeholders and developments in producer responsibility policies (Guo et al. 2017).

Most of the classification factors are generally uncertain in their modelling nature and involves 'uncertainty factors' or objectives that need to be determined precisely for an in-depth analysis of the CLSC system and make changes to it for efficiency, maximizing social impact and reduction in environmental impact. These factors cannot be judged or decided by a certain amount of information and contain number of alternatives or possibilities (Liao et al. 2019a). Uncertainty factors are an inherent and inevitable characteristics of a CLSC system which basically influence the management decision-making process and enhance complexity in it. This complexity hinders the transitional shift from an open-end supply chain system to CLSC (Velte and Steinhilper 2016). Various methods and models have been proposed under various circumstances to determine the "uncertainty factors" and have been reported extensively in literature (Coenen et al. 2018) including few scholars giving detailed statistics on these methods (Govindan et al. 2015). The uncertainty factors along with these methods and models proposed, helps in carrying out an analytical study to find solutions to various issues in CLSC system using multidimensional approaches (Peng et al. 2020).

In this paper, the general theoretical concept of CLSC system, various classification factors in CLSC system as given in Shekarian (2019), the major uncertainty problems that are important and hard to analyze in creating a value supply chain (enhancing profit and at the same time maintaining green and sustainability by reusing, repairing and remanufacturing of the end of use/life products) has been discussed in detail. Various modelling methods and strategies for implementing CLSC system are discussed which helps in determining the uncertainty factors precisely and can be analyzed for an optimal operation among all participating stakeholders.

2 Closed-Loop Supply Chain (CLSC) Systems

Recent governments have started putting in pressure on organizations to develop systems that not only concern for the market profit and customer satisfaction but also to collect, recycle and remanufacture the consumer's end of life/used products. Various regulations and directives such as restriction on hazardous substances (ROHS) and Waste electrical electronics equipments (WEEE) have been initiated to be adopted by the organizations which realize the need for an improved supply chain management system combining the role of forward supply chains or logistics and reverse logistics (RL) (Pokharel and Mutha 2009; Samuel et al. 2020). Govindan et al. (2015) first made an extensive review on CLSC system considering all literature between the year 2007 and 2013 on various factors in RL for a better implementation strategy of the CLSC. CLSC has also been proved to be a suitable approach in assisting organizations in adopting the above-mentioned directives and regulations (Bhattacharya and Kaur 2015).

As in a CLSC system product return from customers are a major input which generally varies in their size, shape, functionality, quality and volume based on the type of industries, the type (or quality) of products received at the collection centres that varies with associated uncertainties. A pre-sorting procedure before the remanufacturing or processing of the returned products is required in order to divert the products having low residual values to recycling facilities from the collection centres. Thus lesser number of returns getting shipped to the Inspection and Refurbishment centres (IRC) facilities for processing, results in lesser carbon emission and cost-saving in transportations enhancing efficiency of the recycling process (Schultmann et al. 2003). Quick sorting is another diversion concept that analyses the number of work cycles the returned product has gone through, providing information on re-manufacturability even before its disassembly operation (Zikopoulos and Tagaras 2008). Pre-sorting at the customer zone is another way (Samuel et al. 2020) for early diversion of the returns with low remanufacturing potential from the value recovery system thus reducing emissions and transportation costs. Pre-sorting helps in reducing the size and number of the reprocessing facilities required in a CLSC network and has the potential to eliminate the variability in return qualities and make strategic plans for long-term choices.

Pollution and carbon emissions can be seen very often in big cities like Delhi in India and Beijing in China which blocks the visibility and is considered a serious threat for various lung and respiratory diseases in humans (Samuel et al. 2020). With the rise in industries and one way supply chain systems the emission has been increasing at an alarming rate due to which policymakers have switched to regulations and carbon emission policies. Emission policies that have been implemented recently in industries can be classified as Carbon cap (CC), Carbon tax (CT) and Carbon cap and trade (CCT). CC is basically a policy which allows a certain maximum amount of carbon emission and restricts beyond that in industries. CT is basically the amount of tax to be paid by the industries for each unit of carbon emission. CCT is a trading system where the industries emitting lesser amount of carbon than the cap limit has extra credits which can be sold out to industries having higher carbon emission than the cap limit, thus maintaining a sustainable atmosphere and opportunities for industries in making extra profit. This profit-making policy also inspire organizations to develop new innovative ways of tackling carbon emission and implement green technologies into their operations. There have been numerous researches in incorporating emission strategies in CLSC system during the stages of manufacturing, transportation, recycling and warehousing (Chen et al. 2017; Xu et al. 2017; Tsao et al. 2017).

The major objective is to incorporate these pre-sorting mechanism and carbon emission policies into the framework of CLSC system which basically is associated with numerous uncertain factors and modelling strategic alternatives. The basic goal of a CLSC system is for an optimized and improved modelling procedure of these uncertainty factors so that the industry advances to a more sustainable, green and profit-making one.

3 Categories/Groups of Factors Affecting the Evaluation of a CLSC System

In the paper (Shekarian 2019), the author analyzed the literature and after screening each paper based on the target and object of the models, the affecting factors of a Game Theory (GT) based model are categorized under certain classes or groups given as economic, environmental, incentive mechanism, financial and single groups. In this section, the first four major groups will be discussed in details. Single groups include all the affecting factors such as E-tail, inventory, quality, information systems, trade-in, cannibalisation and 3Ds such as deterioration, depreciation and defectives, the detailed discussion of which is beyond the scope of the paper.

In the **Economics** group, all the economic factors affecting a CLSC decision-making process are included such as discounts given on the remanufactured product, cost of the remanufacturing, advertisements and services. These subcategories are described in detail as below:

Advertisement Advertisement can be generally implemented by one party or more such as the manufacturer and the retailers in the forward/selling or in reverse operations. Advertisements generally contribute in increasing the sales of the products and return volumes after usage, by manipulating the mental state of its customers towards a more sustainable operation. It plays a vital role in enhancing their returning operation performance, for demand growth (Johnson 2013), supply chain member's profit and pricing and enhancing acceptance probability of the remanufactured products by the customers (Sabharwal and Garg 2013). Though advertising can only be observed in forward supply chain in our day to day life, Jena et al. (2017) stressed upon the role of advertisement in enhancing reverse value chain by persuading customers to return their end of use products for remanufacturing. Advertisement operation can be further subdivided in a forward direction, reverse direction and in both directions.

Discounts Discount is a factor which motivates the customers to purchase manufacturer's new or remanufactured products and improve sustainability in the entire value chain of the products. In the reverse supply chain, discount plays a major role in elevating the sales of a remanufactured product by motivating the customers for their willingness to pay (WTP) for it at a lower rate than its corresponding new product. Apart from the remanufacturing, various other forms of discounts are also included in factors like product quantity, product demand, trade-in, profit functions, different channels and reverse supply chain operations.

Examples can be given of the discount offers by the Prebate recycling program for the customer's willingness to return the empty cartridges as suggested by Lexmark.

Services Services are generally provided by the manufacturers or retailers and has characteristics such as repairing/maintaining the products during customer use, providing information on products to the customers, etc. The different forms of services that are generally discussed in literature include reverse channel services, competition between manufacturers in providing the best services (Wu 2012), cost of services provided by retailers (He et al. 2016; Kong et al. 2017; Zhang et al. 2015) and the customer services for random demands (Zhu and Yu 2018).

Cost Cost is an important factor considered for ensuring a healthy and a sustainable supply chain management in both directions of a CLSC system. Some of the costs associated in both directions of a manufacturing firm include the design cost, manufacturing cost, collection cost, remanufacturing cost and recycling cost. Few forms of costs discussed in the literature as the target of their research include cost of technology (Modak et al. 2018), Research and development cost (Sun et al. 2018), manufacturing costs (Atasu and Subramanian 2012), remanufacturing costs (Ferguson and Toktay 2006), costs in producing interchangeable products (Wu 2013), costs of processes like collection, recycling and disposal (Atasu et al. 2013; Jacobs and Subramanian 2012; Kaushal and Nema 2013).

In the **Environmental** group, the factors which are generally included for a sustainable and green CLSC system are Carbon emission, Green activity and impact and regulations.

Carbon emission As already mentioned in Sect. 1, governments have started to take measures on tackling the environmental issues of the industries by implementing carbon policies such as carbon cap (CC), carbon tax (CT), carbon cap and trade (CCT). Industries emit carbon while extracting raw materials (Li et al. 2014; Shu et al. 2018), during the entire lifecycle of a product (He et al. 2016) and after end of the lifecycle while recycling or remanufacturing. These emissions depend on various uncertainty factors in the value chain such as level of optimality in design, material property of the product influencing the production processes and recyclability or re-manufacturability of the products. In CCT, the government allocates emission credits or quotas to the manufacturers which set a carbon emission target to be considered during manufacturing/remanufacturing (Hu et al. 2016). Similarly, CT can be enforced in industries for further reduction in carbon emission. The manufacturers in order to fulfil these carbon reduction targets can improve on its green and low-carbon emission technologies (Li et al. 2017). In literature, these environmental issues have been discussed considering different targets and approaches such as CCT systems, policy established by the government, emission quantity and emission reduction.

Green activity CLSC systems can be made green and activities be conducted to spread awareness among the consumers about the positive impact of returning their used products and better usage of the products during the lifecycle. CLSC consumers can be classified based on their green activity into primary consumers and green consumers. The primary consumers basically are concerned with the pricing of remanufactured products which they demand to be lower than the corresponding new product based on their affordability. The green consumers, on the other hand, want the remanufactured products to function well throughout their lifecycle period and are concerned about issues in sustainable consumption and environmental impacts. Few strategies and initiative taken in the CLSC system for maximizing the number of returns from the customers under green activity programs are reverse logistics, recovery, recycling, sorting, refurbishing, campaigning for awareness in recycling policies, training programs of employees and monetary incentives (De Giovanni 2014; De Giovanni et al. 2016; De Giovanni and Zaccour 2014). Green activities are concerned with identifying green customers and investment on returning the used products, recycling them or remanufacture.

Impact and regulations In this section, various environmental impact during the lifecycle of a product and the regulations and policies implemented by the governments worldwide are studied and how are they connected. The channels or the objective of the study are different in different literatures. Yenipazarli (2016) observed the dependency of environmental degradation and Environmental impact (EI) on the number of manufactured products. Xiong et al. (2016) analyzed the effect of implementation of the remanufacturing operation on the EI discount. Similarly, EI was studied in a supply chain loop in terms of the quantity of discarded products in landfills in Shi and Min (2013). Miao et al. (2017) took the life cycle assessment of products in order to calculate the total impact on the environment. The effect of the different marketing channels and strategies of the remanufactured products on environmental performance was studied (Yan et al. 2015).

In the **financial** group, all the fluctuation in the cost market, disruptions, profitable risks associated with organizations, franchises have taken by the retailer and patent licensing etc. are covered.

Disruptions Disruptions in the market can occur due to natural or global causes that cannot be prevented. It makes the entire labour cost and remanufacturing profit volatile. Disruption basically includes fluctuation in the order quantity, price and demand (Huang and Wang 2018; Xu et al. 2016), cost of remanufacturing (Han et al. 2016, 2017; Huang and Wang 2017b; Wu and Kao 2018) in a CLSC system. Disruptions can be totally uncontrollable as can be observed during the great earthquake in Japan in 2011 which disrupted supplies in international and national companies and also market demand.

Licensing Licensing is a payment method of the third-party/distributors or the remanufacturers to the manufacturers for carrying out their operations legally which are generally in the form of patent fee (Sun et al. 2018; Yan et al. 2018; Zhang and Ren 2016; Zhang et al. 2018a; Huang and Wang 2017a; Liu et al. 2018; Zou et al. 2016). Patent fee can be of two types which are the fixed fee and the royalty (Hong et al. 2017). In fixed payment, the distributers and the remanufacturers are required to pay the total sum in one transaction and the manufacturers provide them with licenses. On the other hand, in case of royalty, the distributors or remanufacturer tries to make the payments in instalments or paying the sum per unit of products manufactured. Organization profit and performance of the CLSC systems are enhanced and improved by such methods of licensing.

Risks Numerous risks are associated with the CLSC system which needs to be resolved for the common benefit of all stakeholders and the (re)manufacturers. These risks come in the form of market demand, raw material acquisition, remanufacturing quantity, other supply chain risks and the attitude of the stakeholders towards risk handling (Han et al. 2016; He 2017; Wilhite et al. 2014).

Franchise Franchise is a common practice where the manufacturer charges a certain sum that is to be paid by the collector entities or the retailers under contract to carry out their businesses of providing services and products to the consumers. The lump-sum side payment is decided based on the number of order quantities in the profit-making function (of the retailers and manufacturers), received by the retailers (Zhang et al. 2014).

In the **Incentive mechanism** group, various factors associated while designing for the CLSC system includes subsidy from the government, reward-penalty mechanism and two-part tariff contract (TPTC) concept.

Subsidy Subsidy is the support system for the players in the CLSC system or the stakeholders to motivate them in their green/sustainable activities and enhance performance. It is generally provided by the policymakers or the government to the manufacturers, retailers, consumers, remanufacturers, collection centres to encourage recycling of the used products (He et al. 2016; Ma et al. 2018; Zhou et al. 2017), improve demand conditions or increasing consumption, developing the circular economy, curb pollution and carbon emission reduction (Li et al. 2014; Esmaeili et al. 2016; Wan 2018). The categories of subsidies that are generally widely discussed in the literature are manufacturing/remanufacturing subsidies, trade-in subsidy, unit carbon subsidy and recycling subsidy.

Reward penalty When the incentive mechanism or subsidy is accompanied by a penalty triggering a two-sided effect it is called the Reward penalty (RP) mechanism. The RP mechanisms are generally implemented as given in the literature in the processes concerning collection, carbon emission reduction, consumers and remanufacturing. In the CLSC models concerning the collection of used products, RP mechanism can be implemented by setting targets of their collection capacity or rates to ensure more involvement in the recycling process. In the models concerning carbon emission, RP mechanism can be implemented by imposing carbon taxes on the manufacturers as penalty (Yenipazarli 2016; Zhang et al. 2018b). Similarly, RP mechanism can also be implemented in remanufacturing activities and in consumers (Kaushal and Nema 2013; Zhang et al. 2018a; Esmaeili et al. 2016).

Two-part tariff contract (TPTC) A contract by which the retailers are usually paid a certain sum by the manufacturers to encourage collection and recycling activities in their operation. In some cases as in the centralized system, both the manufacturers and the retailers are encouraged to involve in such activities and schemes. A CLSC system considering the original manufacturers, third-party manufacturers, and two tariff parts was proposed in Zhang and Ren (2016). Dobos et al. (2013), Hong et al. (2015) and Shi et al. (2016) considered TPTC in their CLSC system provided to the retailers by the manufacturers to encourage activities related to collection and recycling of products.

4 Methods and Modelling Strategies Considering the Uncertainty Problems in a CLSC System

As already described in the previous sections, various economic, environmental, financial and other factors play a major role in proper functioning of the CLSC system bringing in sustainability into it. All these factors together influence the modelling strategies of the CLSC system which can be classified into four main uncertainty types as given in Peng et al. (2020): Parameter uncertainty, background uncertainty, structural uncertainty of the CLSC models and result uncertainty of the CLSC models (Liao et al. 2019b).

Parameter uncertainty basically includes the quantity, time and quality of the returned products, demand of the recycled products, cost of designing and planning in the CLSC system. *Background uncertainty* generally includes the unknown and ever-changing states of politics, society, ecology, economy and technology such as legislation, changing weather conditions and behaviour of the market demand (Lehr et al. 2013; Besiou and Wassenhove 2016; Chen et al. 2015; Huang et al. 2009; Ruimin et al. 2016). *Structural uncertainties* generally refers to the factors that arise out of human errors, actor ignorance or disagreement. The major uncertainty factor in this category, include prediction of the 'best fit' combination of the CLSC options and are generally treated as fuzzy or stochastic variables. *Result uncertainty* can also be termed as prediction error difference between the real world situations and the outcome of the predicted model. Various tradeoffs involved among the participating players, inconsistency with the cumulative uncertainty of the system model and involvement of multiple CLSC decision areas on the CLSC model results in complications in the solution of the CLSC system.

Classification of methods to incorporate uncertainty factors in CLSC system After identifying all the uncertainties in the CLSC system, these are incorporated into certain methods for data uncertainty evaluation. The methods for this can be generally classified into three types which are the stochastic method (considering probabilistic distribution, chance constraint), fuzzy logic method and the interval programming method (interval values of uncertain parameters).

4.1 Fuzzy Method

In a fuzzy method, the uncertainty factors are basically evaluated by the managers, empirical observation or intuition and are regarded as fuzzy variables or number. The constraints can be included in a fuzzy set and the degree of satisfaction of these constraints is called the membership function through some quantity of dis-satisfaction is also allowed. The intersection of the fuzzy sets gives the membership function and optimal fuzzy decision. Fuzzy method provides a suitable approach for dealing with probabilities and expressions. Many researchers have employed fuzzy methods to carry out their optimization problem such as finding the optimal remanufacturing cost and analysis in CLSC system (Liao et al. 2018; Liu 2002, 2006). Supplier selection and order allocation in a CLSC system were evaluated using a multi-objective model in a fuzzy environment by Moghaddam (2015). Soleimani et al. (2017) solved the supplier selection problem considering opportunity constraints and a fuzzy method in a supply chain system.

4.2 Stochastic Method

Stochastic method has a wide application in all aspects of the CLSC and RL systems today and is originated with the introduction of linear and non-linear programming methods. Optimal order quantities; optimal remanufacturing strategies based on the uncertain demands; optimal location search; lead times in an inventory; quality, quantity and time of the returned products; analyzing degradation and deterioration of the returned products for inventory model are few of the uncertain objective variable that are modelled using the stochastic method (Liberatore 1979; Sphicas 1982; Friedman 1984; Goyal 2007; Ilic et al. 2009; Chang et al. 2001; Cai et al. 2013).

4.3 Interval Method

In many cases, it is not feasible to employ the stochastic and fuzzy method due to non-availability of patterns in a locally determined data, involvement of subjective data and non-availability of sufficient data. Interval methods prove to be the best methods to be employed in these circumstances which basically sets an interval between the ranges of variability in the uncertain parameters and can be denoted by upper and lower bounds. Shevtshenko and Wang (2009) used an intelligent and robust decision support system applying modal interval to design uncertainties in the management of the CLSC system.

Modelling methods In order to carry out various optimization strategies in the CLSC system for the best economic and environmental advantages, mathematical models are constructed with proper mathematical inferences. The application of these mathematical models or modelling methods can vary with different situations for various uncertainty problems in the RL/CLSC system. Peng et al. (2020) had arranged all the modelling methods based on the techniques used in the reviewed papers for a CLSC system into categories of main modelling methods and other modelling methods.

The main modelling methods which are often used extensively in literature and can be applied to larger variants of uncertainty problems. These are: (1) Linear and mixed-integer programming, (2) Non-linear programming methods, (3) Convex and Concave programming, (4) Dynamic Programming, (5) Queuing model, (6) Markov decision process, (7) Graph theory, (8) Game theory, (9) Fuzzy logic, (10) Simulation modelling, (11) Multi-criteria decision-making method and (12) Modelling of conceptual and descriptive types. Other modelling methods include: (1) Artificial neural network, (2) Dynamic regression model, (3) Robust Bayesian belief networks with interval probabilities, (4) Engineering and economics and technology, (5) Input–output analysis and Laplace transform, (6) Production frontier theory and (7) Institutional theory neighbourhood rough set method.

From the reviewed articles, it has been observed that the frequency of use of the linear and non-linear mixed-integer programming methods is the highest as compared to other modelling methods. This implies the robustness and dominance of these two models in solving various uncertainty problems with different complex situations. It has also been observed that almost all co-ordination and pricing problems are modelled analyzed through game theory while for the decision-making problems, apart from the planning and designing can be solved through multi-criteria decision-making method. Simulation modelling is also widely used for different issues by considering several weights given to various uncertainty problems.

Solution techniques Modelling methods and solution techniques share a close dependency and are inter-related. Different solution approaches can be employed while modelling uncertainty problems using linear and non-linear mixed-integer programming modelling methods such as simulation techniques and meta-heuristic algorithms.

Few of the solution technique approaches employed in the literature conducted by [uncertainty] are Analytical or precise methods (AOP), General algebraic modelling systems (GAMS), Multi-objective solution approaches (MOM), Average approximation techniques and methods, Heuristic and meta-heuristic algorithms and simulation technology. Among these, MOM and AOP have been used extensively in the literature. AOP is used as a problem solver yielding complex solutions and are limited in their use for large scale uncertainty problems. In this regard, heuristic and meta-heuristic algorithms are generally used for large scale problems. MOM techniques are generally used for appropriate inter-disciplinary uncertainty problems and include Analytic network process (ANP), Analytic hierarchy process (AHP) and Techniques for order performance by similarity to ideal solution (TOPSIS). For real and practical situations, more powerful techniques can be used such as simulation technology and software.

5 Future Scope

In the economic group of the affecting factors in the evaluation of the CLSC system, advertisements are generally seen to be associated with the forward supply chain system for the sales of the new product rather than in the reverse chain. Involving advertisement activities for remanufactured products and in motivating consumers to return their used products will help gain more sustainability. Studying the effect on discounts under various stochastic real conditions is another future work to be carried out. Though a number of models have evolved in due course of time to incorporate the environmental impact of the products in a CLSC system, the research is still in its infancy period and more stress has to be given on the consumer side rather than on the producer side. Consumers need to be well informed through advertisements about the ill effect of product dumping and emission on environmental degradation and should be motivated for recycling and remanufacturing. Subsidy is a powerful tool that can be provided to the remanufacturers for more returns and reduced prices of the remanufactured products. In the licensing subcategory, investigation regarding the threshold of the fixed fee on the optimal licensing strategy can be conducted as future work. Studies regarding reference quality or the threshold quality level of the product which seems to be influencing both the prices and the sales of the product need to be carried out as future work.

As already discussed, stochastic and fuzzy modelling approaches require a large number of empirical data to be precisely predicted and optimized while the interval modelling approach is based on a range of probabilities within upper and lower bounds. The optimal range can be obtained at the intersection range field but while zooming in the optimal range, there still remains the need to employ deterministic optimization model to find the expected solution. With advancements in the programming and optimization techniques including non-linear programming methods (Özceylan et al. 2014), new pathways have opened for researchers to analyze for a deeper connection between participants/stakeholders and solving complex problems. Also, new innovative modelling methods apart from the traditional ones can provide researchers to search for newer and efficient ideas (Bottani et al. 2015; Azadeh et al. 2016).

6 Conclusion

From the above discussion, it is clear that in order to obtain a green and sustainable CLSC system all the classification factors must be integrated through various uncertainty and modelling methods and solution techniques. The major common objective is to motivate consumer to return their products after end of use/life and recycling/remanufacturing of the products. All stakeholders in a CLSC system must be aware of the environmental impact of a product lifecycle and must be tackled with proper and effective policies/regulations by the government, manufacturers, remanufacturer and retailers. Since the real world is a complex mechanism, different uncertainty and modelling methods/techniques (stochastic, fuzzy and interval) are used in complete harmony for an efficient and environmental friendly CLSC system.

References

Asim, Z., Jalil, S. A., & Javaid, S. (2019). An uncertain model for integrated production-transportation closed-loop supply chain network with cost reliability. *Sustainable Production and Consumption, 17,* 298–310. https://doi.org/10.1016/j.spc.2018.11.010.

Atasu, A., & Subramanian, R. (2012). Extended producer responsibility for e-waste: Individual or collective producer responsibility? *Production and Operations Management, 21,* 1042–1059.

Atasu, A., Toktay, L. B., & Van Wassenhove, L. N. (2013). How collection cost structure drives a manufacturer's reverse channel choice. *Production and Operations Management, 22,* 1089–1102.

Azadeh, A., Zarrin, M., & Salehi, N. (2016). Supplier selection in closed loop supply chain by an integrated simulation-Taguchi-DEA approach. *Journal of Enterprise Information Management.*

Besiou, M., & Wassenhove, L. N. (2016). Closed-loop supply chains for photovoltaic panels: A case-based approach. *Journal of Industrial Ecology, 20,* 929–937.

Bhattacharya, R., & Kaur, A. (2015). Allocation of external returns of different quality grades to multiple stages of a closed loop supply chain. *Journal of Manufacturing Systems, 37,* 692–702.

Bottani, E., Montanari, R., Rinaldi, M., & Vignali, G. (2015). Modeling and multi-objective optimization of closed loop supply chains. *Computers & Industrial Engineering, 87,* 328–342.

Cai, X., Feng, Y., Li, Y., & Shi, D. (2013). Optimal pricing policy for a deteriorating product by dynamic tracking control. *International Journal of Production Research, 51,* 2491–2504.

Chang, H.-J., Hung, C.-H., & Dye, C.-Y. (2001). An inventory model for deteriorating items with linear trend demand under the condition of permissible delay in payments. *Production Planning & Control, 12,* 274–282.

Chen, W., Kucukyazici, B., Verter, V., & Jesús Sáenz, M. (2015). Supply chain design for unlocking the value of remanufacturing under uncertainty. *European Journal of Operational Research, 247,* 804–819. https://doi.org/10.1016/j.ejor.2015.06.062.

Chen, Y.-W., Wang, L.-C., Wang, A., & Chen, T.-L. (2017). A particle swarm approach for optimizing a multi-stage closed loop supply chain for the solar cell industry. *Robotics and Computer-Integrated Manufacturing, 43,* 111–123.

Coenen, J., van der Heijden, R. E. C. M., & van Riel, A. C. R. (2018). Understanding approaches to complexity and uncertainty in closed-loop supply chain management: Past findings and future directions. *Journal of Cleaner Production, 201,* 1–13. https://doi.org/10.1016/j.jclepro.2018.07.216.

De Giovanni, P. (2014). Environmental collaboration in a closed-loop supply chain with a reverse revenue sharing contract. *Annals of Operations Research, 220,* 135–157.

De Giovanni, P., & Zaccour, G. (2014). A two-period game of a closed-loop supply chain. *European Journal of Operational Research, 232,* 22–40.

De Giovanni, P., Reddy, P. V., & Zaccour, G. (2016). Incentive strategies for an optimal recovery program in a closed loop supply chain. *European Journal of Operational Research, 249,* 605–617.

Dobos, I., Gobsch, B., Pakhomova, N., Pishchulov, G., & Richter, K. (2013). Design of contract parameters in a closed-loop supply chain. *Central European Journal of Operations Research, 21,* 713–727.

Dowlatshahi, S. (2000). Developing a theory of reverse logistics. *INFORMS Journal on Applied Analytics, 30,* 143–155. https://doi.org/10.1287/inte.30.3.143.11670.

Esmaeili, M., Allameh, G., & Tajvidi, T. (2016). Using game theory for analysing pricing models in closed-loop supply chain from short- and long-term perspectives. *International Journal of Production Research, 54,* 2152–2169.

Ferguson, M. E., & Toktay, L. B. (2006). The effect of competition on recovery strategies. *Production and Operations Management, 15,* 351–368.

Friedman, M. F. (1984). Balancing stocks and backlogs with stochastic lead times: A special case. *Computers & Operations Research, 11,* 233–239.

Georgiadis, P., & Besiou, M. (2010). Environmental and economical sustainability of WEEE closed-loop supply chains with recycling: A system dynamics analysis. *International Journal of Advanced Manufacturing Technology, 47,* 475–493. https://doi.org/10.1007/s00170-009-2362-7.

Govindan, K., Soleimani, H., & Kannan, D. (2015). Reverse logistics and closed-loop supply chain: A comprehensive review to explore the future. *European Journal of Operational Research, 240,* 603–626. https://doi.org/10.1016/j.ejor.2014.07.012.

Govindan, K., Mina, H., Esmaeili, A., & Gholami-Zanjani, S. M. (2020). An integrated hybrid approach for circular supplier selection and closed loop supply chain network design under uncertainty. *Journal of Cleaner Production, 242,* 118317. https://doi.org/10.1016/j.jclepro.2019.118317.

Goyal, S. K. (2007). Optimal ordering policies for a retailer in a supply chain with up-stream and down-stream trade credits. *Journal of the Operational Research Society, 58,* 1252–1255.

Guide, V. D. R., Jr., & Van Wassenhove, L. N. (2009). OR FORUM—The evolution of closed-loop supply chain research. *Operations Research, 57,* 10–18.

Guo, S., Shen, B., Choi, T.-M., & Jung, S. (2017). A review on supply chain contracts in reverse logistics: Supply chain structures and channel leaderships. *Journal of Cleaner Production, 144,* 387–402.

Han, X., Wu, H., Yang, Q., & Shang, J. (2016). Reverse channel selection under remanufacturing risks: Balancing profitability and robustness. *International Journal of Production Economics, 182,* 63–72.

Han, X., Wu, H., Yang, Q., & Shang, J. (2017). Collection channel and production decisions in a closed-loop supply chain with remanufacturing cost disruption. *International Journal of Production Research, 55,* 1147–1167.

He, Y. (2017). Supply risk sharing in a closed-loop supply chain. *International Journal of Production Economics, 183,* 39–52.

He, R., Xiong, Y., & Lin, Z. (2016). Carbon emissions in a dual channel closed loop supply chain: The impact of consumer free riding behavior. *Journal of Cleaner Production, 134,* 384–394.

Homrich, A. S., Galvao, G., Abadia, L. G., & Carvalho, M. M. (2018). The circular economy umbrella: Trends and gaps on integrating pathways. *Journal of Cleaner Production, 175,* 525–543.

Hong, X., Xu, L., Du, P., & Wang, W. (2015). Joint advertising, pricing and collection decisions in a closed-loop supply chain. *International Journal of Production Economics, 167,* 12–22.

Hong, X., Govindan, K., Xu, L., & Du, P. (2017). Quantity and collection decisions in a closed-loop supply chain with technology licensing. *European Journal of Operational Research, 256,* 820–829.

Hu, D., Xiao, C., & Chen, X. (2016). Carbon quotas, subsidies and engineering machinery remanufacturing. *Frontiers of Engineering Management, 3,* 50–58.

Huang, Y., & Wang, Z. (2017a). Closed-loop supply chain models with product take-back and hybrid remanufacturing under technology licensing. *Journal of Cleaner Production, 142,* 3917–3927.

Huang, Y., & Wang, Z. (2017b). Dual-recycling channel decision in a closed-loop supply chain with cost disruptions. *Sustainability, 9,* 2004–2031.

Huang, Y., & Wang, Z. (2018). Demand disruptions, pricing and production decisions in a closed-loop supply chain with technology licensing. *Journal of Cleaner Production, 191,* 248–260.

Huang, X. Y., Yan, N. N., & Qiu, R. Z. (2009). Dynamic models of closed-loop supply chain and robust H_∞ control strategies. *Computer Integrated Manufacturing Systems, 47,* 2279–2300.

Ilic, A., Ng, J. W. P., Bowman, P., & Staake, T. (2009). The value of RFID for RTI management. *Electronic Markets, 19,* 125–135.

Jacobs, B. W., & Subramanian, R. (2012). Sharing responsibility for product recovery across the supply chain. *Production and Operations Management, 21,* 85–100.

Jena, S. K., Sarmah, S., & Sarin, S. C. (2017). Joint-advertising for collection of returned products in a closed-loop supply chain under uncertain environment. *Computers & Industrial Engineering, 113,* 305–322.

Johnson, J. P. (2013). Targeted advertising and advertising avoidance. *The Rand Journal of Economics, 44,* 128–144.

Kaushal, R. K., & Nema, A. K. (2013). Strategic analysis of computer waste management options: Game-theoretic approach. *Journal of Environmental Engineering, 139,* 241–249.

Kong, L., Liu, Z., Pan, Y., Xie, J., & Yang, G. (2017). Pricing and service decision of dual-channel operations in an O2O closed-loop supply chain. *Industrial Management & Data Systems, 117,* 1567–1588.

Lehr, C. B., Thun, J. H., & Milling, P. M. (2013). From waste to value—A system dynamics model for strategic decision-making in closed-loop supply chains. *International Journal of Production Research, 51,* 4105–4116.

Li, J., Du, W., Yang, F., & Hua, G. (2014). The carbon subsidy analysis in remanufacturing closed-loop supply chain. *Sustainability, 6,* 3861–3877.

Li, H., Wang, C., Shang, M., & Ou, W. (2017). Pricing, carbon emission reduction, low-carbon promotion and returning decision in a closed-loop supply chain under vertical and horizontal cooperation. *International Journal of Environmental Research and Public Health, 14,* 1332–1357.

Liao, H., Deng, Q., Wang, Y., Guo, S., & Ren, Q. (2018). An environmental benefits and costs assessment model for remanufacturing process under quality uncertainty. *Journal of Cleaner Production, 178,* 45–58.

Liao, H., Deng, Q., & Shen, N. (2019a). Optimal remanufacture-up-to strategy with uncertainties in acquisition quality, quantity, and market demand. *Journal of Cleaner Production, 206,* 987–1003.

Liao, H., Shi, Y., Liu, X., Shen, N., & Deng, Q. (2019b). A non-probabilistic model of carbon footprints in remanufacture under multiple uncertainties. *Journal of Cleaner Production, 211,* 1127–1140.

Liberatore, M. J. (1979). Technical note—The EOQ model under stochastic lead time. *Operations Research, 27,* 391–396. https://doi.org/10.1287/opre.27.2.391.

Liu, B. (2002). Uncertain programming: Optimization theory in uncertain environments. In *IEEE International Conference on Fuzzy Systems FUZZ-IEEE.*

Liu, Y. K. (2006). Convergent results about the use of fuzzy simulation in fuzzy optimization problems. *IEEE Transactions on Fuzzy Systems, 14,* 295–304.

Liu, H., Lei, M., Huang, T., & Leong, G. K. (2018). Refurbishing authorization strategy in the secondary market for electrical and electronic products. *International Journal of Production Economics, 195,* 198–209.

Ma, H., & Li, X. (2018). Closed-loop supply chain network design for hazardous products with uncertain demands and returns. *Applied Soft Computing, 68,* 889–899. https://doi.org/10.1016/j.asoc.2017.10.027.

Ma, Z.-J., Hu, S., Dai, Y., & Ye, Y.-S. (2018). Pay-as-you-throw versus recycling fund system in closed-loop supply chains with alliance recycling. *International Transactions in Operational Research, 25,* 1811–1829.

Merli, R., Preziosi, M., & Acampora, A. (2018). How do scholars approach the circular economy? A systematic literature review. *Journal of Cleaner Production, 178,* 703–722.

Miao, Z., Fu, K., Xia, Z., & Wang, Y. (2017). Models for closed-loop supply chain with trade-ins. *Omega, 66,* 308–326.

Modak, N. M., Modak, N., Panda, S., & Sana, S. S. (2018). Analyzing structure of two-echelon closed-loop supply chain for pricing, quality and recycling management. *Journal of Cleaner Production, 171,* 512–528.

Moghaddam, K. S. (2015). Fuzzy multi-objective model for supplier selection and order allocation in reverse logistics systems under supply and demand uncertainty. *Expert Systems with Applications, 42,* 6237–6254.

Mohammed, F., Selim, S. Z., Hassan, A., & Syed, M. N. (2017). Multi-period planning of closed-loop supply chain with carbon policies under uncertainty. *Transportation Research Part D: Transport and Environment, 51,* 146–172. https://doi.org/10.1016/j.trd.2016.10.033.

Neto, J. Q. F., Walther, G., Bloemhof, J., van Nunen, J. A. E. E., & Spengler, T. (2010). From closed-loop to sustainable supply chains: The WEEE case. *International Journal of Production Research, 48,* 4463–4481. https://doi.org/10.1080/00207540902906151.

Özceylan, E., Paksoy, T., & Bektaş, T. (2014). Modeling and optimizing the integrated problem of closed-loop supply chain network design and disassembly line balancing. *Transportation Research Part E, 61,* 142–164.

Peng, H., Shen, N., Liao, H., Xue, H., & Wang, Q. (2020). Uncertainty factors, methods and solutions of a closed loop supply chain—A review for current situation and future prospects. *Journal of Cleaner Production.* https://doi.org/10.1016/j.jclepro.2020.120032.

Pokharel, S., & Mutha, A. (2009). Perspectives in reverse logistics: A review. *Resources, Conservation and Recycling, 53*(175), 182.

Prieto-Sandoval, V., Jaca, C., & Ormazabal, M. (2018). Towards a consensus on the circular economy. *Journal of Cleaner Production, 179,* 605–615.

Rogers, D. S., & Tibben-Lembke, R. S. (1999). *Going backwards: Reverse logistics trends and practices.* Pittsburgh, PA: Reverse Logistics Executive Council.

Ruimin, M. A., Yao, L., Jin, M., Ren, P., & Zhihan, L. V. (2016). Robust environmental closed-loop supply chain design under uncertainty. *Chaos, Solitons & Fractals, 89,* 195–202.

Sabharwal, S., & Garg, S. (2013). Determining cost effectiveness index of remanufacturing: A graph theoretic approach. *International Journal of Production Economics, 144*, 521–532.

Samuel, C. N., Venkatadri, U., Diallo, C., & Khatab, A. (2020). Robust closed loop supply chain design with presorting, return quality and carbon emission considerations. *Journal of Cleaner Production*.

Schultmann, F., Engels, B., & Rentz, O. (2003). Closed-loop supply chains for spent batteries. *Interfaces, 33*, 57–71.

Shekarian, E. (2019). A review of factors affecting closed-loop supply chain models. *Journal of Cleaner Production*. https://doi.org/10.1016/j.jclepro.2019.119823.

Shevtshenko, E., & Wang, Y. (2009). Decision support under uncertainties based on robust Bayesian networks in reverse logistics management. *International Journal of Computer Applications in Technology, 36*, 247–258.

Shi, W., & Min, K. J. (2013). A study of product weight and collection rate in closed-loop supply chains with recycling. *IEEE Transactions on Engineering Management, 60*, 409–423.

Shi, Z., Wang, N., Jia, T., & Chen, H. (2016). Reverse revenue sharing contract versus two-part tariff contract under a closed-loop supply chain system. *Mathematical Problems in Engineering, 2016*, 1–15.

Shu, T., Huang, C., Chen, S., Wang, S., & Lai, K. (2018). Trade-old-for-remanufactured closed-loop supply chains with carbon tax and government subsidies. *Sustainability, 10*, 3935–3959.

Soleimani, H., Govindan, K., Saghafi, H., & Jafari, H. (2017). Fuzzy multi-objective sustainable and green closed-loop supply chain network design. *Computers & Industrial Engineering, 109*, 191–203.

Souza, G. C. (2013). Closed-loop supply chains: A critical review, and future research. *Decision Sciences, 44*, 7–38.

Sphicas, G. P. (1982). On the solution of an inventory model with variable lead times. *Operations Research, 30*, 404–410.

Sun, L., Zhang, L., & Li, Y. (2018). Sustainable decisions on product upgrade confrontations with remanufacturing operations. *Sustainability, 10*, 4090–4105.

Tsao, Y.-C., Linh, V.-T., Lu, J.-C., & Yu, V. (2017). A supply chain network with product remanufacturing and carbon emission considerations: A two-phase design. *Journal of Intelligent Manufacturing*, 1–13.

Velte, C. J., & Steinhilper, R. (2016). Complexity in a circular economy: A need for rethinking complexity management strategies.

Wan, N. (2018). The impacts of low carbon subsidy, collection mode, and power structure on a closed-loop supply chain. *Journal of Renewable and Sustainable Energy, 10*(065904), 1–24.

Wilhite, A., Burns, L., Patnayakuni, R., & Tseng, F. (2014). Military supply chains and closed-loop systems: Resource allocation and incentives in supply sourcing and supply chain design. *International Journal of Production Research, 52*, 1926–1939.

Wu, C.-H. (2012). Price and service competition between new and remanufactured products in a two-echelon supply chain. *International Journal of Production Economics, 140*, 496–507.

Wu, C.-H. (2013). OEM product design in a price competition with remanufactured product. *Omega, 41*, 287–298.

Wu, C.-H., & Kao, Y.-J. (2018). Cooperation regarding technology development in a closed-loop supply chain. *European Journal of Operational Research, 267*, 523–539.

Xiong, Y., Zhao, Q., & Zhou, Y. (2016). Manufacturer-remanufacturing vs supplier-remanufacturing in a closed-loop supply chain. *International Journal of Production Economics, 176*, 21–28.

Xu, X., Liu, J., Li, Y., & Zhou, X. (2016). Pricing strategy of closed-loop supply chain under disruptions. *Filomat, 30*, 4059–4072.

Xu, Z., Pokharel, S., Elomri, A., & Mutlu, F. (2017). Emission policies and their analysis for the design of hybrid and dedicated closed-loop supply chains. *Journal of Cleaner Production, 142*, 4152–4168.

Yan, W., Xiong, Y., Xiong, Z., & Guo, N. (2015). Bricks vs. clicks: Which is better for marketing remanufactured products? *European Journal of Operational Research, 242*, 434–444.

Yan, W., Li, H., Chai, J., Qian, Z., & Chen, H. (2018). Owning or outsourcing? Strategic choice on take-back operations for third-party remanufacturing. *Sustainability, 10,* 151–168.

Yenipazarli, A. (2016). Managing new and remanufactured products to mitigate environmental damage under emissions regulation. *European Journal of Operational Research, 249,* 117–130.

Zhang, C.-T., & Ren, M.-L. (2016). Closed-loop supply chain coordination strategy for the remanufacture of patented products under competitive demand. *Applied Mathematical Modelling, 40,* 6243–6255.

Zhang, P., Xiong, Y., Xiong, Z., & Yan, W. (2014). Designing contracts for a closed-loop supply chain under information asymmetry. *Operations Research Letters, 42,* 150–155.

Zhang, Z.-Z., Wang, Z.-J., & Liu, L.-W. (2015). Retail services and pricing decisions in a closed-loop supply chain with remanufacturing. *Sustainability, 7,* 2373–2396.

Zhang, D., Zhang, X., Shi, B., Cao, J., & Zhou, G. (2018a). Collection and remanufacturing of waste products under patent protection and government regulation. *Sustainability, 10,* 1402–1423.

Zhang, X., Su, Y., & Yuan, X. (2018b). Government reward-penalty mechanism in closed-loop supply chain based on dynamics game theory. *Discrete Dynamics in Nature and Society, 2018,* 1–10.

Zhou, W., Zheng, Y., & Huang, W. (2017). Competitive advantage of qualified WEEE recyclers through EPR legislation. *European Journal of Operational Research, 257,* 641–655.

Zhu, X., & Yu, L. (2018). Differential pricing decision and coordination of green electronic products from the perspective of service heterogeneity. *Applied Sciences, 8,* 1207–1225.

Zikopoulos, C., & Tagaras, G. (2008). On the attractiveness of sorting before disassembly in remanufacturing. *IIE Transactions, 40,* 313–323.

Zou, Z.-B., Wang, J.-J., Deng, G.-S., & Chen, H. (2016). Third-party remanufacturing mode selection: Outsourcing or authorization? *Transportation Research Part E: Logistics and Transportation Review, 87,* 1–19.

Project-Based Supply Chain Intelligence and Digital Fabrication for a Sustainable Film Industry

Jennifer Loy

Abstract There is a societal, political and legislative drive towards greater transparency and accountability for environmental, social and economic sustainability in business. There is also the recognition of the complexities involved in monitoring operations in manufacturing and construction industries, and in particular where those are project-based. Developments in digital technology provide new ways of tracking elements within supply chain operations, as well as new communication platforms, fabrication techniques and data analysis capabilities. This chapter provides an argument for the need for transition research in project-based manufacturing and construction industries, with film production as an example, to provide the conceptualisation of supply chain constructs to respond to contemporary aspirations for society. These should exploit the opportunities digital technologies provide in the development of supply chain intelligence, whilst working with cultural and practical complexities within the industry, to create the paradigm shifts necessary for an effective re-organisation of practice. This chapter discusses the issues and challenges involved in adapting project-based supply chains for sustainability legislative oversight, and the development of sustainable supply chain practices enabled by digital technology, to propose alternative practices for project-based industries overall.

Keywords Digital technologies · Innovation · Transition research · 3D printing · Additive manufacturing

1 Introduction

The growth in scale of productions in the film industry over the last decade belies early predictions that home viewing would usurp movie theatres, reducing the demand for large-scale blockbusters. As streaming services, such as Netflix and Amazon Prime, vie for home viewing dominance in 2020, the scale of investment in productions for any media platform continues to rise. The stakes are higher, as the costs escalate alongside the scale of the productions, as are the potential profits and losses involved.

J. Loy (✉)
Deakin University, Geelong, Australia
e-mail: Jennifer.Loy@deakin.edu.au

K. Kumar and J. P Davim (eds.), *Supply Chain Intelligence*, Management and Industrial Engineering, https://doi.org/10.1007/978-3-030-46425-7_3

Achieving the look and structure of a set, and preparing it to a timetable, where the availability of the site (even within a studio lot) is limited, means that the priority is the timing of the outcome and its effect. Managing the operations involved is driven by those two priorities. However, like all industries, the effects departments in the film industry are subject to the current pressures and legislation to improve their environmental footprint.

Developing sets and props for a major film is a practical challenge in design and making, involving product designers, visual designers and mechanical engineers. Whilst working to a budget should be critical, managing the evolving requirements on a set and allowing the director to capture the desired shot mean that sets and props—and budgets—are frequently reworked or added to during the shoot. Whilst the visual effects (post-production computer graphics) has developed alongside the advances in digital technologies. Practical effects, including special effects, remain predominantly craft-based activities. The ability to make changes to sets and props at short notice is essential and therefore the materials needed on-site and their management is difficult to predict. Supply chain optimisation, whilst of concern to all organisations involved with a flow of materials and manufacture, is secondary to meeting the needs on-site during a shoot.

In addition to the world-building achieved on set, the equipment that supports the activities onset, such as cherry-pickers for overhead shots, have to be booked in advance and brought onto set at considerable expense, at times only in case they are needed on the day. This is because of the window of opportunity available to capture a scene. Assembling a large crew and actors, and setting up filming for a day for a particular shoot, is difficult to replicate if there needs to be a delay. This may be because of the availability of the site, the weather, the availability of personnel or the cost of extending filming for technical reasons, or because of the constraints of the overall filming schedule.

For the production managers in any organisation, supply chain optimisation is critical to ensure predictability, reliability and manage costs. A film production, in essence, is an extreme version of 'just-in-time' manufacturing. It is a heightened example of project-based logistics, where the project team is generally assembled only for the project, the time frames are limited and the needs inconsistent. The essentials of supply chain logistics are the same for all operations: the right product, at the right place, at the right time, at the right price, in the right amount, however, as with all project-based industries, the management of the flow of materials is subject to the very different demands in service of a project. This makes it difficult to predict and manage. For oversight agencies, such as environmental protection agencies, it is also difficult to monitor and evaluate practice.

There is a lack of research into supply chain optimisation in the film industry in particular, but there are lessons to be learnt from other project-based industries. For example, there are similarities between film production logistics and humanitarian logistics, and whilst there are also key differences, there are experiences that can be transferred from humanitarian logistics and applied to film production logistics. There has been considerable research on humanitarian logistics recently, and there have been developments in practice in the sector as a whole that could inform the film industry. In addition, there is a growing body of research on the integration of technology into supply chains to enable more accurate tracking, monitoring, data analysis and visualisation. This digital integration allows for the potential of supply chain intelligence to be explored, to help the management of operations within the film industry itself, and environmental agencies to monitor practices in the sector. The maturing of the cumulative digital technologies that constituted the digital revolution over the last twenty years, offers ways to rethink systems and new ways of working for a more transparent, sustainable industry sector.

The development of digital fabrication technologies, such as additive manufacturing known as 3D printing, provides new ways of creating bespoke products on demand. The film industry is a good example of where this could create a paradigm shift. Rather than simply incremental change, digital fabrication based on digital inventory could, potentially, rework supply systems for the sector, because of the bespoke nature of the objects required onset and the ability for digital inventories to be developed and shared across projects. This could lead the sector into more long-term planning strategies, and collaborative, more sustainable practices. This chapter discusses these issues and opportunities, and highlights the constraints and possibilities that digital technology could bring to supply chain innovation.

2 Environmental Regulations

All industries are subject to environmental regulations. These regulations and standards cover the impact of operations on the environment, such as greenhouse gas emissions, water pollution, and landfill. As an example, companies have to report their greenhouse (GHG) gas data annually. This data allows companies to track their own emissions, identify where emissions may need to be cut, minimise energy costs and ultimately save money, whilst providing a responsible service to the community. City councils, state oversight and governments can use the data to identify where comparable facilities are producing disproportionately high emissions and can use the data overall to develop emissions control strategies. The Environmental Protection Agency (EPA) in the US requires that organisations report their GHG gas data if they produce over 25,000 metric tons of CO_2 per year, or where the CO_2 in products produced by a manufacturer in a year if released, would amount to over 25,000 metric tons. Alternatively, if the facility uses underground gas injection methods and requires more than 25,000 metric tons of CO_2, the company must also report to the EPA. There is a National Emission standard in the US for hazardous air pollutants and there is also a permit requirement system for companies discharging pollutants into water sources. The EPA states that industrial wastewater permits limit the discharge and conditions of release of water pollutants into the watercourse, based on the type of facility.

Environmental regulations vary throughout the world, but there is a recognition by all agencies of the importance of updating the regulations to respond to the particular current context. In the state of Queensland in Australia, for example, the

Environmental Protection Regulation 2019 was brought in as a "new, contemporary risk-based approach" to 'reflect current waste management standards, practices and technologies" (QLD Gov. DES 2019). This schedule organised waste into two major categories: high risk and moderate risk, with the remainder classified as lowest risk, non-regulated waste. For companies, the waste generator and waste receiver must both keep records of what constitutes the waste. This is organised as under:

- Organic material processing
- Waste disposal
- Reprocessing: Mechanical, thermal waste, other
- Resource recovery (e.g. metals).

Environmental and social impact management and tracking involve being aware of the sourcing and processing of materials, and treatment of waste throughout the lifecycle of products as they are manufactured, and, ideally, as the customer unpacks and uses them, and then on to the end-of-life for that product, including any reuse or recycling or material reclamation. This is not a straight forward activity, and relies on self-reporting, based on a company's own research data, if that company has had an interest in tracking it as part of lifecycle assessment and analysis. However, this reporting is increasingly part of the triple bottom line accounting for companies in the twenty-first century, as illustrated, for example, in the recommendations by the Australian Federal Government, Department of Business (2019). Triple bottom line accounting is where a company defines its success not only on its financial bottom line, but also on environmental accountability and social justice. This approach was proposed more than twenty years ago, as critical to the development of sustainable practices, based on the definition of sustainability as referring to environmental, economic and societal responsibilities.

> To refuse the challenge implied by the triple bottom line is to risk extinction. Nor are these simply issues for major transnational corporations: They will increasingly be forced to pass the pressure on down their supply chains, to smaller suppliers and contractors. These changes flow from a profound reshaping of society's expectations and, as a result, of the local and global markets businesses serve. Anyone who has worked in this area for any time knows there are waves of change. (Elkington 1997, p. 2)

The sustainability imperative has grown (e.g. Hawkins et al. 2010; McDonough and Braungart 2013), as has the acceptance of triple bottom line accounting. The 2007 publication of the IPPC Fourth Assessment Report (AR4) by the United Nations Intergovernmental Panel on Climate Change (IPCC), added to expectations placed on companies, with climate change science formally acknowledged and a call for a more integrated response:

> Designing sustainable countermeasures for addressing global warming requires an approach that unifies the various aspects of climate change, including impact assessment, prediction, mitigation and adaptation measures, policy issues, and social issues. It is essential to attack the problems from a wide range of viewpoints from different academic fields, including natural science, engineering, agriculture, economics, and political science. (Hiramatsu et al. 2008, p. 201)

The report recommended the development of "a new academic discipline of sustainability science, which must adopt a comprehensive and holistic approach to identification of problems and perspectives involving the sustainability of global, social, and human systems." (Hiramatsu et al. 2008, p. 202). The intention was to work towards creating systems that were as close to closed ecological systems as possible. In nature, a closed ecological system is an interconnected system of biological organisms with no additional inputs and no extraneous outputs. An industrial ecological system aims to mimic as closely as possible a closed ecological system. This involves the management of the following:

- The input of materials
- The amount of energy used at each stage
- The output of waste at each stage and the reuse of waste in other products
- The reclamation of materials
- The impact of actions on the environment
- The impact on workers, consumers, society and individual communities.

Whilst this establishes a comprehensive ideal for the different stakeholders involved, it is a significant challenge. Tackling the transition to closed ecological systems thinking requires addressing issues individually, whilst working towards these aspirations for the benefit of society and the environment in the twenty-first century.

3 Creating a Paradigm Shift

> In well-ordered situations, with clear goals and standards and stable conditions, the pursuit of perfection makes sense. But not when we face complex and chaotic conditions, with standards that keep evolving. (Klein 2017, p. 155)

One of the challenges to establishing sustainability standards is tracking the information required by environmental agencies for monitoring the full supply chain of manufactured products. The difficulty is the practicalities involved in the tracking. Currently, the majority of data available to agencies is based on submissions made by the companies themselves. Comprehensive lifecycle assessment is expensive and difficult to organise as information has to be collected and analysed all through the supply chain, from minor suppliers to operations within the factory, to tracking the use and disposal of the product, to reuse and recycling efforts, the reclamation of materials, through to the end-of-life. External monitoring conducted by environmental agencies usually depends on responding to particular problems highlighted in the data, such as where an environmental agency observes a significant difference between the data being provided by two similar facilities. Any larger intervention requires considerable manpower and longitudinal studies on the operations of a particular company.

As difficult as this activity is for the manufacturers themselves, and those involved in their regulation, at least the organisation of the majority of manufactures is based on a predictable input of materials and flow through the factory and a measurable output to users. The unpredictability in the current system for conventional manufacturing is in the use of the product, and its disposal. The introduction of legislation in Europe to help address this situation required the return products to their point of origin for material reclamation. The regulation is called Extended Producer Responsibility (EPR). Under this legislation, manufactures are responsible for the end-of-life of their products. This means manufacturers are liable for both the financial and physical recycling of their products. Technically, EPR makes the manufacturer responsible for all stages of the supply chain, but it is the end-of-life element that is most visible and receives the most focus. In Europe, this applies to all electronic goods, batteries, packaging and vehicles.

For automobile companies, in particular, this legislation has created very different ways of working. Vehicles are large consumer products, traditionally not easily recycled. However, the legislation has created a change in practice for the industry. Cars are now returned to the manufacturers at the end-of-life. These cars are therefore redesigned to provide a degree of disassembly, with detachable parts. The remainder is typically fed through an industrial scale shredder. The materials are then separated, for example by filtering out the metals, and as much as possible recycled. The company is financially responsible for any parts sent to landfill and therefore is motivated to reduce that amount as much as possible through design and its recycling facilities and capabilities.

In other industries, companies have either individually, or as a consortium addressed environmental pressures specific to their sector in response to EPR. For example, the plastics industry is facing considerable environmental pressure. Although their products have been commercially produced for less than seventy years the scale of the impact of plastic waste across the world is overwhelming. According to Geyer et al. an estimated:

> 8300 million metric tons (Mt) of virgin plastics have been produced to date. As of 2015, approximately 6300 Mt of plastic waste had been generated, around 9% of which had been recycled, 12% was incinerated, and 79% was accumulated in land-fills or the natural environment. If current production and waste management trends continue, roughly 12,000 Mt of plastic waste will be in landfills or in the natural environment by 2050. (2017, p. 1)

As part of a response by the plastics industry, eight major companies, including Dow Chemical Company, set up a collaboration called the National Polystyrene Recycling company to research the end-of-life of polystyrene, identified as one of the most harmful. However, plastic products overall provide a universal challenge in relation to their environmental impact, particularly with regards to the use of plastics in packaging. The challenge faced is significant as the durability of plastics, a valued characteristic in product design, also makes it resistant to degradation:

> The only way to permanently eliminate plastic waste is by destructive thermal treatment, such as combustion or pyrolysis. Thus, near-permanent contamination of the natural environment with plastic waste is a growing concern. Plastic debris has been found in all major ocean

basins … with an estimated 4 to 12 million metric tons (Mt) of plastic waste generated on land entering the marine environment in 2010 alone … Contamination of freshwater systems and terrestrial habitats is also increasingly reported … as is environmental contamination with synthetic fibers … Plastic waste is now so ubiquitous in the environment that it has been suggested as a geological indicator of the proposed Anthropocene era. (Geyer et al. 2017, p. 1)

Company consortium solutions to addressing the aspects of the global problem of plastic production, use, recycling and disposal will, therefore, need to be supported by even broader, large-scale legislative changes brought about by governments working together to change not only the manufacture and disposal of these products, but also the consumer habits fuelling the use of these products. This requires a shift towards dematerialising the economy, and changing consumer expectations and behaviours. Changing the paradigm from product supply to product service systems' supply supports that shift in thinking as well as practice. Product service systems is a term that refers to products provided within an organised system (Hawkins et al. 1999). This approach has developed over the last two decades as companies work towards sustainability targets, and reduce their environmental footprint for a dematerialised economy whilst retaining their economic viability. An example of this is Interface Carpets. This company used to manufacture traditional rolls of carpet for commercial properties. It shifted to a product service system approach specifically to improve its environmental credentials. The results demonstrate the ability of a company to benefit financially as well as ethically from a shift to sustainable design. The company now makes carpet tiles, rather than rolls of carpet, and instead of selling carpet to commercial companies, they lease it. That is, they provide the 'service' of carpeting for the company for a set period of time. During the contract, they replace any worn or damaged tiles individually as needed. The tiles are designed to be taken apart and the materials recycled. This example demonstrates that industries can make that sustainability paradigm shift from traditional operations to a product service system approach if the drive to do so is strong enough. This can either come from within a company, such as with the example of Interface Carpets where the CEO Ray Anderson changed his thinking, or where legislation dictates the need for change.

The United States (US) was slow to adopt Extended Producer Responsibility (Nash and Bosso 2013) beyond several state-wide initiatives (particularly in California, Maine and Vermont), but in 2019, federal legislation was introduced that enforced responsibility for the end-of-life arrangements and costs for product packaging, such as drink bottles and cups, food containers, wrappers and lightweight plastic bags (Carra and Meng 2019). National deposit schemes were introduced for all glass, plastic and aluminium beverage containers, and the legislation required the installation of reverse vending machines for their collection. Styrofoam was banned and all plastics were required to have clear labelling on their classification for recycling and their method of disposal. Rules on the recyclability of many plastic products were enforced. However, unlike in the vehicles example, the products involved were less visible, and the legislation, therefore, less able to be enforced. To date, however, in the US vehicle manufacturers are not yet subject to Extended Producer Responsibility, other than with regards to a National Vehicle Mercury Switch recovery program.

The construction industry has recycling issues related to those in the film industry. Currently, on-site construction dominates prefabrication, and at the end-of-life of a structure, as they tend to be currently designed and built, the costs and practicalities of deconstructing and reclaiming materials are prohibitive. For buildings in Europe, Extended Producer Responsibility is in place, but difficult to enforce for whole buildings. However, EPR policies do allow for some waste to be diverted from landfill, as part of practices where the whole of a building cannot realistically be recycled. According to Guggemos and Horvath (2003, p. 1076), legislation to increase the use of recycled materials "may save energy, reduce virgin material use, and prevent pollution. Economic instruments can also be used to promote EPR for buildings, whilst information instruments are not as effective." Adherence to this legislation will continue to be difficult to independently track and enforce unless the way the construction industry operates has a paradigm shift. Buildings need to be designed and constructed with end-of-life in mind, and supply chain operations need to be more transparent. The problems for the construction industry in adopting digital technology and new ways of operating in a digital age to aid this conversion are similar to those for the film industry, as it would affect all operations. Not least, workforce training. Addressing how to the change established practices, cultural norms and entrenched attitudes in the construction industry, and in the craft-based prop and set departments of the film industry, is arguably more complex than for manufacturers where cultural expectations can be built in-house.

The twenty-first century has seen a shift in thinking about sustainable practices, and also about the development of the urban landscape towards digitally enabled, smart city environments. Communication technologies are changing the way communities interact and organise within an urban environment. In response, designers and architects have begun proposing the development of more connected, flexible environments that challenge conventional ideas on the length of time a structure should stand, what it should facilitate, and how it can be adapted or dismantled in response to changing needs. This responds to the idea that spaces are increasingly temporary, and should, therefore, be easily converted or even moved. Essentially, those advocating for a 'pop-up' city approach (Beekmans and de Boer 2014) argue that traditionally built urban landscapes are outdated. Spaces need to be adaptable to changing requirements, and assumptions about use and the materials invested in the building for the length of time it will be suitable for its intended purpose need to be questioned. There is also an argument for the removal of the conventional split between the planning of urban spaces for professional or leisure activities. The Dutch design group Droog, for example, worked with architects Diller, Scofido and Renfro and locals in a US suburb to encourage home businesses through architectural interventions. The team created open house prototypes demonstrating how communities could be redesigned for more equitable, sustainable systems in the twenty-first century (Beekmans and de Boer 2014, p. 58).

> The way contemporary cities are made no longer fits the dynamic of the age. Why does a society in which people become exponentially more flexible and mobile by the minute fail to make cities more adaptable to change? Cities are intense sites of activity and innovation, and yet most urban planning departments appear to be stuck in a postwar mode fails to address the needs of new activities and new users. (Beekmans and de Boer 2014, p. 16)

According to these initiatives, the construction of buildings needs to be rethought at this time to change so that it is possible for communities to respond to the changing needs of their members. Short-term use spaces should be integrated with structured but adaptable spaces. For a long-term view of the supply chain innovations needed to provide for these changes and the development of an idealistic shared, flexible inventory to service these operations, digital technology needs to be integrated into practice. Digital technologies allow for communication platforms, and connected, customised products (Novak and Loy 2018) that can be tracked and reused as part of a design for disassembly strategy (Walker 2013). These could potentially revolutionise the building industry if it is possible to make such a wholesale change to established practices and attitudes. This would require enforced legislation and financial incentives by governments. Similarly, in the film industry, a shift towards more collaborative practices across projects, an increased emphasis on building structures for the time they were meant to be used for, and the reclamation of material and component parts would change practices. Similarly to the Interface example, whilst this may seem unachievable and expensive, it may prove economic over time. There would need to be a recognition that this would change what was built, who owned it and the organisation of how it was dismantled and materials reclaimed. This would require complex transition research that is currently not in evidence, a growing area of research discussed later in the chapter. What is missing is the incentive. This may come from sustainability legislation, such as Extended Producer Responsibility, being enforced in industries involved in built environment projects such as the film industry, or it may come from communities or companies themselves as the digitally enabled society matures. According to Beekmans and de Boer (2014, p. 17), current trends will force communities to change the way they think about building, with the intention of "building better, more resilient urban areas; urban areas teeming with temporary pop-up initiatives are prompting cities themselves to become more responsible and adaptive to the needs and desires of locals and visitors alike."

4 Project-Based Supply Chain Logistics

In the decade starting 2020, a more rigorous approach to waste stream management should become a priority for manufacturers responding to legislative drivers. As discussed, this should extend to the construction industry, creating a shift towards more adaptive design and design for disassembly, however difficult it is to create such a paradigm shift. The construction industry is, to an extent, classified as project-based in comparison to conventional manufacturing companies. The film industry, however, provides a more extreme example of project-based working. The manufacturing sector can work towards evolving a 'lean' forward supply chain, and associated forward logistics systems (Martin 2002), creating efficiencies for the benefit of the company and its customers, and utilising assembly line efficiency strategies, such as Just-In-Time (JIT) for the supply of materials and components during assembly (Giordano and Schiraldi 2013). Innovating project-based supply chains is more problematic, not least because their temporary nature means the actors tend to be smaller companies. This is critical because, despite the perceived agility and responsiveness of smaller companies, research has shown that larger companies are in a better position to explore innovating business practices and supply chain organisation and management because they are able to rely on a longer time frame for their operations (Kasperson 2001). Corporations cited as having introduced complex industrial ecology approaches include Apple Computers, General Motors, IBM, Hewlett-Packard, Motorola and Xerox (Martin 2002, p. 26). Smaller companies, particularly those operating in a project-based industry, where a temporary supply chain is brought together on the basis of a single project, are unlikely to be in a position to innovate practice in response to environmental concerns. For this reason, these collaborations of companies tend to be slow at innovating and responding to environmental pressures, yet are critical to widespread adoption of environmental policies: "The key challenge in driving higher overall corporate environmental performance will be one of forging consistency—how to bring along the laggards and late adopters more rapidly to the pace-setting standard of industrial leaders." (Kasperson 2001, p. 12).

In comparison to conventional manufacturing supply chain operations, the logistics for project-based operations are complex because they are less predictable. Material, product and equipment sources may vary significantly with each project. To meet its project requirements, many production companies tend to rely on repeat cycles of temporary supply chain organisations, formed at relatively short notice. Time is, therefore, lacking for the development of the trust between stakeholders needed to build innovations for the long-term development of industry sector practices, in response to the opportunities provided by digital technologies, or the pressures of the growing sustainability imperative. The stresses of a condensed time frame for project delivery, and the bespoke nature of the props and construction required, driven by management teams often brought specifically for a project, mean it is likely there is little appetite for pre-empting regulatory change by collaboratively updating industry practices. Yet, irrespective of the priorities for individual project teams in the film and television industries, just as other industries have to respond to the sustainability imperatives driving legislative and practice change, it is arguably time for the sector as a whole to objectively review its practices in order to align with the sustainability aspirations of society at this time. This is in part because of the growth in scale of productions over the last two decades, and therefore the growth in the environmental footprint of those productions. It is also, however, because emerging digital technologies relevant to the operations of the industry could create smart supply chain innovations of benefit to the sector long-term. Digital technology enables new ways of working, supporting the integration of supply chain intelligence through digital platform development, digital communication and fabrication. If the industry was forced to rethink its practices a whole, then the tools are now available to provide radically new ways of working that could prove stabilising in the long term, though are not without risk in the short term. A platform to build these practices could be emerging as key leaders in the industry have begun establishing successive film productions, planned over years, that would allow for transition research to occur that would inform practice across the sector.

5 Shared Inventory and Supply Chain Intelligence

> It is one thing to note the need for corporations to be proactive in environmental performance, quite another to deliver on the goal with consistency and effectiveness. (Kasperson 2001, p. 12)

As noted, there is currently a dearth of research on supply chain operations for the film and television industries, but there has been a growing interest in research into humanitarian supply chain operations. Whilst the drivers for the organisations working in the two sectors are different, there are similarities in experience that enable developments in humanitarian logistics to be considered with respect to film industry supply chain intelligence. In both, the sectors consisted of multiple organisations working independently of each other, dependent on short-term relationships established for projects of varying size and requirements that were difficult to anticipate. In both, time and reliability were critical. In both, material and products' supply may involve goods being brought onto a site from overseas as well as from local providers. In both, the priority was not the development of an organisational structure, but the delivery of the requirements on a short-term basis. One of the challenges this creates is a lack of focus in developing professional personnel to oversee supply chain operations on a long-term basis or to compare practices across companies. This means the industry itself is unlikely to drive research into changing practices. This is illustrated by the example given by Sigala and Wakolbinger (2019, p. 56) of research into the recovery phase of equipment and materials in humanitarian projects:

> The recovery phase seems not to have a clear agenda for the private sector yet and reverse logistics is not that developed in humanitarian operations, but the role of the private sector is also important at this stage. As mentioned by a private sector expert "we try to be involved in the recovery of the cold storage equipment back into Europe, but it's only now becoming part of the logistics agenda for these organizations that you could consider doing reverse logistics as well. So, it's a much undefined area and it's still a long way off".

Where initiatives to improve humanitarian logistics can inform the film industry, is in the shift from a fragmented sector to a more collaborative one. Research highlighted the importance of developing greater transparency in humanitarian logistics, and the development of a supply chain that was more predictable and manageable (e.g. in relation to product and material supplies, transportation and storage, Loy and Tatham 2016). One of the results was a collaborative initiative where disparate organisation shared centralised warehousing and created inventory that could be directed wherever it was required at relatively short notice. This reduced the risks in single sourcing for individual organisations, and improved capacity building and tracking for the sector as a whole. This allows the sector a degree of stability that enables the development of supply chain intelligence strategies, such as the use of shared digital platforms and the gathering and analysis of tracking data for the overall sector. This in turn could lead to increased efficiencies and reduce the stress associated with each task.

For the film industry, one of the major challenges to developing sector-wide sustainability operations and relevant, practical legislation is its fragmentation. Whilst individual production companies can operate responsibly, collaboratively, or with long-term strategies in mind, based on supplying the unpredictable requirements for temporary projects, the opportunities or incentives for innovation are reduced. If the approach taken in the humanitarian logistics sector was adapted for the film industry, this could provide the basis for more sector-wide development. The nature of project-based operations creates a disjointed flow of goods and services that is difficult to innovate. The development of industry-specific online platforms to support closed-loop systems, and the introduction of the widespread use of digital tracking and data analysis to evaluate and rationalise the distribution and collection of materials and products, could provide a basis for the development of supply chain intelligence. As the cost of sensors, etc., drops, and the capabilities of machine learning rise, this should add value and stability to the industry, not least in monitoring the use and recycling of materials for feedback to the environmental protection agencies, to help improve the overall understanding of the impact of the industry and where its operations could be improved. As the scale of projects increases, this will become increasingly more important. Acting proactively to anticipate legislative changes, particularly through investment in research in this field, would support the industry in the future.

6 3D Printing and Changing Practice in the Film Industry

In addition to creating the practice of a shared inventory in film production, supported by digital communication technology, recent developments in digital fabrication are allowing for a shift towards a digital inventory that could also revolutionise supply chain operations. Additive manufacturing, commonly termed 3D printing, refers to a range of technologies where an object is created layer by layer from a 3D model built in computer-aided design software. Although they share this approach to building structures, the fabrication technologies are very different. For example, some are resin-based, some extrusion-based and others powder-based, with a laser used to fuse points on the 3D object in layers. A wide range of materials are available, from biomaterials to carbon fibre, to titanium. Each process has different constraints and designing for each requires different strategies (Diegel et al. 2019). In common, however, is the ability for a 3D model designed for 3D printing to be stored in a digital inventory, ready to be printed as required. Working this way is still relatively uneconomic for mass manufacturing industries as the unique selling point of 3D printing is that each print can be different from the previous one. In conventional manufacturing, parts are predominantly created using moulds or other forms of tooling. As a result, the economics of mass production practices apply. The greater the number of identical parts produced from a single mould or tool, the more economic. In 3D printing this is not the case, as it costs the same to print each design, even if they are different. However, the cost of mass production using 3D printing is currently higher than the cost of conventional mass-produced articles. Therefore, the technology is only worthwhile if the geometry of the part has to change for each print. For this reason, 3D printing has been adopted extensively by the medical device industry, where bespoke manufacturing to the requirements of an individual human body are needed. The hearing aid industry provides a good example. In 2020, hearing aids are predominantly produced using 3D printing as each design matches the scan of an individual's ear. The film industry requires one-off, bespoke products for temporary use. The parts have to be functional to the specifications required for filming, for example, a suit of armour may be required to be very light or flexible for a particular scene, or equally, it may need to be produced in sections that breakaway, and have weight to them. A helmet or face mask will need to fit the individual actor, and an animatronic model of an alien may need complex geometries, not possible with traditional fabrication techniques. As an example of an industry that additive manufacturing is most suitable for, the film industry meets the brief.

With respect to supply chain innovation, the integration of 3D printing into practice allows for the development of a digital inventory. This means that parts can be replicated from digital files. Storing parts between productions are difficult and costly. Space has to be found, and the reality is that future productions in a series, even if planned, may be cancelled due to factors beyond the production company's control. Equally, storing objects on the off-chance they may be used in the future is problematic because they are not designed and built for durability, and because the requirements for each individual shoot tend to be different. Even the use of different actors, with different measurements, will mean a stored article may be of no use in a succeeding production. By retaining digital files, the company is able to retain the design investment in a piece, and then reprint the article as it is, or adapt it to meet the requirements of the new film. As the use of 3D printing in the film industry increases, there should be a development of digital libraries of parts suitable for printing props.

The use of 3D printing in the film industry in 2020 is not widespread, except for visual models or prototyping. Where it has been adopted for end-use parts, there are two factions. The first is in stop motion animation. The example of stop motion animation provides an insight into the workforce issues for studios looking to adopt 3D printing. Stop motion animation is traditionally a craft-based skill. The animators work with their hands to sculpt detailed character studies to tell a story. Each facial expression is sculpted onto the character's face. It is a slow process requiring considerable expertise. This expertise has not been aligned to the skills required for computer modelling. As a result, there can be resistance to the introduction of a digital technology that requires a different skill set, and, at first sight, be seen to eliminate the jobs of studio artists by 3D printing the models in a fraction of the time. In part, these fears are justified. Hand-making skills and 3D digital modelling skills, plus the knowledge of designing for a 3D printing process are not conventionally taught together. In education, this will only change as established academic add to their knowledge with professional development, or are replaced over time with educators self-taught in digital technology. Without the academic leaders to set the dual curriculum, the adoption of the technology will be slow. As Legge (2012, p. 69) observed: "Change can be difficult at the best of times, but when it challenges the status quo, it can be downright frightening."

The second part of the concern of workers is less justified. Henry Selick and Tim Burton's cutting edge stop motion animation, The Nightmare Before Christmas, was made in 1993. It was nominated for an Academy Award for Best Visual Effects—unusual for a stop motion animation—and illustrated best practice at the time. In this film, the heads of the main character, Jack Skeleton, were sculpted by hand. There were 700 heads in total. Each had a different expression. Sixteen years later, in 2009, a US company called Laika Studios,[2] developed a project called Coraline. The studio used a plaster-based 3D printing technology to print the heads. The heads were built using parametric modelling, a technique that allowed the foundations for each model to be retained whilst the details changed. The plaster-based printer allowed numerous heads to be printed simultaneously. If the animators were reproducing the Jack Skeleton character, then printing 700 heads would be a relatively quick activity. The process would justify the concern of the traditional animators that the technology would reduce their numbers. However, as usually happens with new technological capabilities, the demand rises to meet the supply. In this case, for Coraline, the animators 3D printed over 20,000 heads. These prints were made using the original plaster-based technology. The face was bisected into two, with an upper eyebrow section, and lower mouth section, allowing for over 200,000 facial expressions. The prints were white, and the surfaces had to be sanded and painted by hand. Rather than reducing the role of the animators and makers, the technology-enabled an expansion of what was possible. Laika then went on to produce ParaNorman in 2012, printing over 33,000 heads, with separate eyes, allowing for over a million different facial expression combinations. This was followed by Boxtrolls, where an estimated 50,000 heads were printed.

In 2016, Laika upgraded the 3D printing technology it used from plaster-based, to multi-coloured, multi-material, polymer-based. The technology worked much as a 2D printer, depositing droplets of polymer from banks of print heads, but to build three-dimensional models. The result was pre-coloured, multi-material characters with interchangeable parts in their faces, and a multitude of facial expressions. These parts did not need to be finished by hand, but the transition from hand sculpting, through plaster-based 3D printing, to multi-material, multi-coloured printing over seven years would have allowed animators to upgrade their skills to include digital modelling. The techniques the animators evolved, and a character called The Moon Beast was developed, made up of 3D printed shells attached to a central armature. It was the first fully 3D printed puppet that Laika had undertaken, built of 130 separate 3D printed pieces. In addition, Laika has added visual effects to their animations, integrating two traditionally disparate disciplines. The company received a Scientific and Technology Oscar for its work and has been presented with multiple animation awards.

Whilst this example illustrates how embracing digital technology into a traditional operation, can facilitate new ways of thinking and working in an organisation, rather than simply replicating existing practices, the second use of 3D printing illustrates the potential of digital fabrication technology integration to impact supply chain logistics. In this example, 3D printing is used for building props and sets for the film industry. Its impact on the logistical models commonly employed in the film industry can be significant, demonstrating the potential of digital technologies to disrupt, rather than enhance existing business and logistics practices. Legacy Effects is a US company that specialises in the film industry. Their history is in special effects, and their work in mechatronics and puppeteering was widely recognised as cutting edge in the industry. Legacy Effects initially bought a dual filament printer, with the intention of creating prototypes. However, over the last decade, the company has realised the potential of the technology and now produces props and costumes, for example for Iron Man and Robo Cop, integrating the technology fully into its working practice. As with Laika, the technology is not used to replicate conventional practice, but instead it enables the company to expand its offerings. This means, for example, that costumes can be adapted to have different sections designed to collapse on impact or be replaced to give the character a different look for a different scene. The quantity of parts being produced for film has increased with 3D printing technology, allowing for different activities during the shoot, and the costumes can be fitted to the actors, using 3D scanning techniques.

For the supply chain, shifting to 3D printing for this particular company allows it to bring the manufacture of parts in-house. Although materials still have to be supplied, the company can keep an inventory of the raw materials on-site, and convert it into a product on demand. This contraction of the supply chain changes the working practices of the company. It allows the company to manage and monitor material use, making legislative reporting easier. Also, as a digital technology, the history of what was printed and in which materials will be recorded. However, according to Yosofi et al. (2019) even though there has been a growing interest in research on the sustainability of additive manufacturing processes, "there is still a lack of reliable inventory data. This makes it difficult to conduct an accurate life-cycle assessment or sustainability analysis." Where a production company does not have the 3D printers it needs in-house (a metal printer for example in 2020, will cost upwards of $800,000 US), there is a growing number of service bureaus. Working with service bureaus will change the supply chain operations of a company, but not shift the industry towards more distributed manufacturing, or significantly reduce the transport miles for materials and products that negatively contribute to sustainable practices.

> A sustainable supply chain is one that can generate profits over an extended period without harming the social or natural system. (Montabon et al. 2016)

Again, although the supply chain logistics may be different using service bureaus, and whilst not necessarily more efficient as information has t be effectively communicated, parts developed and printed, then transported, it does ensure that accurate records are available and across the industry, rather than on a case by case basis. This should help environmental reporting and the ability for researchers to develop more sustainable industry practices long-term.

There are 3D printing companies making large-scale props that can be used on set, but 3D printing suitable for actual set construction is in its early stages. Whilst there are examples of building practices that can clearly be seen to be relevant to the future of set production in the film industry, such as the work of Dus Architects[3] at the Canal House in Amsterdam, or the MX3D[4] metal bridge, currently developments are not sufficiently progressed for its integration into production practice for film. However, the research work of companies such as the 3D printing specialist architects, Emerging Objects,[5] point to new ways of working with materials for temporary products suitable for set design, for example with their salt-based room structures. For the film industry working to improve its environmental credentials, digital technology, such as 3D printing and digital inventories, allow for new ways of making temporary products that could revolutionise the industry and challenge the dominance of conventional materials.

> The relative advantages and disadvantages of dematerialization, substitution, reuse, material recycling, waste-to-energy, and conversion technologies must be carefully considered to design the best solutions to the environmental challenges posed by the enormous and sustained global growth in plastics production and use. (Geyer et al. 2017)

According to Wang et al. (2016, p. 1) "… emerging technologies (e.g., Internet of Things (IoT)…wireless sensor networks…big data…cloud computing…embedded system…and mobile Internet)…are being introduced into the manufacturing environment, which ushers in a fourth industrial revolution." Ustundag and Cervikcan (2018, p. 5) argue for highlighting the roles of simulation, virtualisation, adaptive robotics and additive manufacturing in Industry 4.0. Connected industrial systems are enabled by the developments in computing power, sensors and actuators, RFID and RTLS and mobile technologies, and activated by real-time data management in integrated business processes, with interoperability, and the ability for an agile response to changing requirements. However, Ustundag and Cervikcan (2018, p. 5) also state that currently "A structured and systematic implementation roadmap for the transformation to Industry 4.0 is still uncertain." In data analytics, for example, identified as a central component of smart supply chains, three types are discussed in research. These are data analytics concerned with optimisation, description and prediction. Wang et al. (2016), argue that although descriptive and predictive data analytics receive far less attention in research than optimisation, the relationship between the three categories should be the subject of greater study, as they should be viewed as phases of supply chain operations, rather than distinct approaches. However, there is a dearth of research into the siloed elements themselves that need to be rectified for credible data analytics to be used in supply chain operations:

> Many review articles point to the widespread use of descriptive analytics, however, with the exception of demand forecasting, case examples showing how predictive data analytics can be used in supply chains are scarce at best. As several authors report that supply chains became more complex and global, the issue of anticipating and preparing for disruptions have become increasingly important (Sheffi 2005). While several scholars have hinted that predictive algorithms can be used in supply chain risk management (Wang et al. 2016; Nguyen et al. 2018) no empirically based contributions exist to date (Brintrup et al. 2019, p.1).

As manufacturing operational approaches, such as Industry 4.0 and its successors, become more prevalent with the growth in digital capabilities, including data analysis, and the reduction in cost of the physical hardware needed, industries will be in a position to implement supply chain intelligence innovations that respond to the sustainability needs of society at the moment, potentially without a negative economic impact. What is required is a change of thinking, and a holistic view of complex interactions and competing demands in each industry, in order to maximise the digital technologies that have only recently become available. This is about educational development and cultural change as much as it is about digital technological innovations.

7 Development of Transition Research

According to Kasperson (2001), corporate culture is the manifestation of the accepted norms, values and beliefs of a corporation. It establishes the assumptions that will be made during the operations of that company, and informs its mission statement, its policies, its organisational structure and rules of behaviour. One of the challenges in developing disruptive change for a company, and more broadly for an industry sector, is the perceived threat to existing practices and hierarchy. However beneficial to society a shift towards sustainable practices maybe, if the changes involved require very different practices, then it may face resistance within individual companies:

> Executives may believe that they want insights and innovations but are most receptive to new ideas that fit with existing practices and maintain predictability. Business organisations treat disruptive insights and innovations with suspicion. Witness the initial hostile reactions to the telephone, to Google's search engine, to VisaCale, to the Xerox 914 copier, and to Xerox's rejection of its own personal computers. All these innovations became highly successful, but corporations initially were suspicious of these technologies and tried to dismiss them. (Klein 2017, p. 154)

If the companies in an industry sector are reluctant to engage with innovations in a particular sphere, such as sustainability; disruptive digital technologies or the development of data-driven supply chain intelligence, one issue in the current research environment is funding support. Industry-driven research is central to the idea of ensuring that academic research is relevant. Yet, if this is the case, there is a danger that industry research will be driven by commercial concerns and short-term thinking. According to Bucolo (2015, p. 151): "Managing these two mindsets in parallel—today's business model and a future opportunity is what firms need to overcome constantly". Transition research is required to help shift project-based industries to more sustainable operational practices, and to optimise their supply chains informed by advances in digital technology, including analytics and machine learning. However, the industry-funding required to prioritise this is unlikely to be available. In addition, a literature review by Wong and Ngai (2019), concluded the focus of supply chain innovation research has been biased towards conventional manufacturing organisations. For research in this area to be more generally applicable, future studies should be from more diverse situations. The lack of project-based supply chain innovation research evidence supports that conclusion.

In translational research terms, supply chain innovations need to be considered in longitudinal studies but also viewed through the results of multi-dimensional and cross-sectional studies (Papachristos 2014). This ensures all stakeholders are considered as part of systems thinking approach:

> Inevitably supply chains being part of a sociotechnical system, they also contribute to its inertia. Therefore a transition towards sustainability requires that the operation of supply chains (micro level) be altered as well towards modes of operation with lower carbon emissions, material disposal and environmental impact. These alternative operational modes are referred to in the literature as "green" supply chain or closed-loop supply chains… The importance of supply chains as integral elements of sociotechnical systems implies that the relevance of their operations and operations strategies for transition processes must be explored. (Papachristos 2014, p.1)

According to Lee et al. (2011), supply chain innovations are designed to improve organisational processes in new ways by using new technologies to address situational uncertainty and changing customer needs. Over the last decade, digital technologies for the collection of data relating to the operations and performance of elements along the supply chain have evolved to a level of sophistication that allows for complex processes to be developed. However, there is currently little evidence that this has been applied to project-based supply chain operations for the reasons given earlier in the chapter. For the adoption of digital technology into the supply chain, there need to be internal or external drivers in the industries concerned, but also there needs to be a focus on transition research that is not solely longitudinal and observational, but also cross-sectional. Ideally, it should also lead to transition interventions research to enable project-based industries to move to more predictable operations. The challenge is that project-based operation, and in particular those concerned with disciplines that have established traditions, including in terms of the skills and expertise of the workers involved, need to be the focus of transition research sensitive to the people impacted by the change. For the opportunities for efficiency, stability and monitoring, particularly for the environmental management for the benefit of society as a whole, to be capitalised on by such industries, the impact on people, and the transition of conventional working practices into the development of a new workforce needs to lead the research. Although Wong and Ngai 2019, p. 159) defined supply chain innovations as one of three key activities: "logistics-oriented, marketing-oriented, and technological development-oriented innovation activities." Ojha et al. (2016) suggested, supply chain innovation was more complex and people-centred, stating that successful supply chain innovation has to be overtly cross-organisational.

They argued that it has to recognise and work with corporate culture as a relational phenomenon, rather than being seen as a mechanistic process.

As supply chain intelligence is developed, facilitated by digital technologies and multi-disciplinary research, it may be possible to develop suitable instruments in the future as data becomes more readily available across industries. However, whilst technology development-oriented innovation focusses on new technical skills and tools, the qualitative study of the experience of stakeholders along the supply chain, including the development of temporary and long-term partnerships and collaborations, needs to be understood. Montabon et al. (2016) argue that in current strategies for creating supply chain innovations that are directed at sustainability imperatives are based instrumental logic. That is they address how supply chain operations can be improved for a company by addressing environmental or social issues, rather than asking how a supply chain can, in itself, become sustainable. They state that instrumental logic in this context has two significant weaknesses:

> First, this logic is backwards looking in that it studies existing unsustainable supply chains to determine what they are doing to become less unsustainable…In extant research, sustainability is mainly addressed from the perspective of what can existing firms do to reduce their harm while maintaining or increasing their profits. Research of this nature cannot lead to truly sustainable supply chains because it addresses trade-offs by prioritizing the profits of existing firms over other sustainability outcomes including the survival of society and the environment. Second, while sustainable supply chain research is ostensibly aimed at the entire chain and all of its stakeholders, the reality is that it is usually conducted from the perspective of a focal firm. This means that research to date has mainly investigated sustainability related performance measures of the focal firm while generally overlooking other members of the chain and the communities in which the supply chain operates. (Montabon et al. 2016)

8 Conclusion

Over the last decade, digital technologies have sufficiently matured to allow supply chain innovations that, applied appropriately in an industry-specific context, can lead to the development of supply chain intelligence. Digital monitoring, data analysis, shared platforms, new ways of interacting enabled by digital technology, digital inventory and collective knowledge banks for the industry, digital fabrication and a shift to distributed manufacturing for project-based work that reduces transport miles, all suggest the potential for disruptive change. To exploit the potential of digital technologies to provide supply chain innovation, there needs to be leadership in the proposal of conceptualisation frameworks for the development of supply chain intelligence in a digital age. This will only be possible with the development of research into supply chain innovation, and in the context of this chapter, in transition research for the development of theory, practical strategies and measurement instruments for project-based, supply chain intelligence systems. Research into administrative innovation and product development may have overshadowed supply chain innovation research in the past, but there now needs to be supply chain innovation research into both incremental and radical sector change with the evolving digital environment. This is both for the future of the industries concerned and in anticipation of increased monitoring required for sustainability legislation. Project-based industries, such as the construction industry, provide extreme use cases suitable for transition research at this time. Film production provides an extreme use case within that subset, suitable for research that will inform project-based industry operations management overall.

Triple bottom line accounting, measuring success in terms of financial, environmental and social metrics, is become increasing integral to business practice. Environmental legislations, such as EPR, have been accepted as standard in many countries, although the pace of acceptance by a government or state administration varies. One of the key challenges for legislative bodies is the monitoring of activities. Lifecycle tracking is currently predominantly based on self-reporting, with agencies assessing the data overall to identify potential problems. As digital technologies are maturing, there are opportunities for companies and oversight organisations to track more accurately supply chain operations. For project-based industries, the complexities in developing and enforcing appropriate legislation are significant. In addition, in industries where there is an existing culture of practice that is based on skills and expertise traditionally unrelated to digital expertise, there are frequently cultural challenges that need to be addressed as part of a shift towards a digitally enabled system. Film sets are temporary products, in a project-based system. As such, film production provides a good case study for transition research into digitally enabled, sustainable practices in a project-based industry. The industry is subject to environmental legislation, but this is likely to be affected by developments in the near future. For example, the rigour of the reclamation process of construction materials after a shoot is inevitably balanced with the need to vacate a film lot, or external site being used for the activity, particularly on films where the exterior sites are essentially on public land. Without a significant external driver, there are likely only to be incremental changes of practice as the industry is based on traditional operations and established skills and expertise, and fragmented by its project-based nature. Whilst currently legislation for the construction industry and the film production industry avoids the wholesale waste material controls imposed on manufacturers, this will inevitably change in the future. The industry is currently experiencing a change in supply chain operations where key players are world-building, planning successive film productions on a massive scale. This provides an opportunity to consider supply chain optimisation as the work shifts from project-based to one with more manageable, predictable requirements. This would allow an opportunity for research on a paradigm shift towards product service system thinking. Transition research on projects such as these could help the industry to move towards greater sustainability in the sector overall.

Leading directors of large-scale productions are committed to world-building on a grand scale over the coming years. Where the planning of a series of mammoth projects over a number of years is assured, production companies are in a position to act as stable, enduring organisations leading the sector. It is, therefore, arguably, the time for governments and film industry associations to invest in transition research to support the development of the production landscape overall. This strategy would support building supply chain intelligence strategies for the industry, based on the digital technology tools that are now available, and on research into transition interventions. There is a need for transition research for film production, as an example of a project-based growing in scale, that could benefit from a holistic response to developments in digital technology for a more sustainable future. The industry has its distinct history and established practices, and these need to be recognised and integrated into transformational strategies. The reality is that at some point legislation, or societal pressures will direct attention to the industry. It would be beneficial if the industry could anticipate the need for more sustainable practices, and make the changes on its own terms.

> I have suggested that we need big, audacious reforms for a reason. Such thinking snaps us out of complacency of the status quo and helps break the habit of thinking 'there is no alternative'. We need ambitious ides for change and the courage to pursue them. (Dunlop 2018, p. 233)

Notes

1. Laika Studios: https://www.laika.com.
2. Legacy Effects: http://www.legacyefx.com.
3. DUS Architects: https://houseofdus.com.
4. MX3D: https://mx3d.com.
5. Emerging Objects: https://www.emergingobjects.com.

Glossary

Supply chain intelligence Refers to a digitally connected supply chain, where data is collected from each stage of the supply chain, and its analysis informs decision making in supply chain operations.

Sustainability The ability to retain economic viability whilst avoiding negative impacts on the environment or society.

3D printing A range of processes that build 3-dimensional objects in layers from computer models, also called additive manufacturingadditive manufacturing.

Digital inventory The virtual storage of part date where the quality and usability of that part is verified.

Industry 4.0 This term refers to a trend towards automation and the use of cyberphysical systems in manufacturing.

References

Australian Government, Department of Business. (2019). Accessed on January 2, 2020. https://www.business.gov.au/Risk-management/Environmental-impact/Environmental-management-and-your-business.

Beekmans, J., & de Boer, J. (2014). *Pop-up city: City making in a fluid world*. Amsterdam: BIS.

Brintrup, A., Pak, J., Ratiney, D., Pearce, T., Wichmann, P., Woodall, P., & McFarlane, D. (2019). Supply chain data analytics for predicting supplier disruptions: A case study in complex asset manufacturing. *International Journal of Production Research*. Available from Taylor & Francis. https://doi.org/10.1080/00207543.2019.1685705.

Bucolo, S. (2015). *Are we there yet?*. Amsterdam: BIS.

Carra, R., & Meng, D. (2019). Federal packaging extended producer responsibility legislation under development. Beveridge & Diamond, August 6. Accessed on December 28, 2019. https://www.bdlaw.com/publications/federal-packaging-extended-producer-responsibility-legislation-under-development/.

Diegel, O., Nordin, A., & Motte, D. (2019). *A practical guide for design for additive manufacturing*. London: Springer.

Dunlop, T. (2018). *The future of everything: Big, audacious ideas for a better world*. Sydney: New South Publishing.

Elkington, J. (1997). *Cannibals with forks*. Oxford: Capstone.

Esenduran, G., Hall, N., & Liu, Z. (2015). Environmental regulation in project-based industries. Wiley Online Library. https://doi.org/10.1002/nav.21625.

Geyer, R., Jambeck, J., & Law, K. (2017). Production, use and the fate of all plastics ever made. *Science Advances, 3*, e1700782. Accessed January 3, 2020, available from https://advances.sciencemag.org/content/advances/3/7/e1700782.full.pdf.

Giordano, F., & Schiraldi, M. (2013). On just-in-time production leveling. In *Operations management*. Intech Open. https://doi.org/10.5772/54994. Accessed on December 29, 2019, available from https://www.intechopen.com/books/operations-management/on-just-in-time-production-leveling.

Guggemos, A., & Horvath, A. (2003). Strategies of extended producer responsibility for buildings. *Journal of Infrastructure Systems, 9*(2), 65). 10.1061/(ASCE)1076-0342(2003).

Gurdus, L. (2018). *Dow chemical CEO: "I've never seen the industry more aligned' around tackling plastic waste*. CNBC. August 2, accessed on December 15, 2020, available from https://www.cnbc.com/2018/08/02/dow-chemical-ceo-on-plastic-waste-problem-industry-initiatives.html.

Hahn, G. (2019). Industry 4.0: A supply chain innovation perspective. *International Journal of Production Research*. https://doi.org/10.1080/00207543.2019.1641642.

Hawkins, P. (2010). *The ecology of commerce*. New York: Harper Business.

Hawkins, P., Lovins, A., & Lovins, H. (1999). *Natural capitalism: The next industrial revolution*. London: Routledge.

Hiramatsu, A., Mimura, N., & Sumi, A. (2008). A mapping of global warming research based on IPCC AR4. *Sustain Science, 3*, 201–213. https://doi.org/10.1007/s11625-008-0058-9.

Kasperson, R. (2001). 'Foreword'. In J. Sarkis (Ed.), *Greener manufacturing and operations: From design to delivery and back* (pp. 11–15). Oxen: Taylor & Francis.

Klein, G. (2017). *Seeing what others don't: The remarkable ways we gain insights*. London: Nicholas Brealey.

Lee, S., Lee, D., & Schneiderjans, M. (2011). Supply chain innovation and organizational performance in the healthcare industry. *International Journal of Operations & Production Management, 31*(11), 1193–1214. https://doi.org/10.1108/01443571111178493.

Legge, K. (2012). *Doing it differently*. Sydney: Place Partners.

Loy, J., & Tatham, P. (2016). Redesigning production systems. In I. Haavisto, G. Kovacs, & K. Spens (Eds.), *Supply chain management for humanitarians: Tools for practice*. London: Kogan Page.

Martin, M. (2002). Implementing the industrial ecology approach with revers logistics. In J. Sarkis (Ed.), *Greener manufacturing and operations: From design to delivery and back* (pp. 24–36). Oxen: Taylor & Francis.

McDonough, W., & Braungart, M. (2013). *The upcycle: Beyond sustainability—Designing for abundance*. New York: North Point Press.

Ming, C. (2018). *Plastic pollutions: Firms and governments are combating millions of tons of waste.* CNBC April 22, accessed on January 10, 2020, https://www.cnbc.com/2018/04/22/plastic-pollution-firms-and-governments-fight-waste.html.

Montabon, F., Pagell, M., & Wu, Z. (2016). Making sustainability sustainable. *Journal of Supply Chain Management, 52*(2).

Nash, J., & Bosso, C. (2013). Extended producer responsibility in the United States: Full speed ahead?' In *Regulatory policy program working paper* RPP-2013-04. Cambridge, MA: Mossavar-Rahmani Center for Business and Government, Harvard Kennedy School, Harvard University.

Novak, J., & Loy, J. (2018). Digital technologies and 4D customised design: Challenging conventions with responsive design. In V. Bryan (Ed.), *Handbook of research on human development in the digital age* (pp. 403–426). IGI Global: Ann Musgrove & Jillian Powers.

Papachristos, G. (2014). Transition inertia due to competition in supply chains with remanufacturing and recycling: A systems dynamics model. *Environmental Innovation and Societal Transitions, 12,* 47–65.

Queensland Government Australia, Department of Environment and Science. (2019). Accessed on December 12, 2019, https://environment.des.qld.gov.au/management/waste/business/classification.

Sigala, I., & Wakolbinger, T. (2019). Outsourcing of humanitarian logistics to commercial logistics service providers: An empirical investigation. *Journal of Humanitarian Logistics and Supply Chain Management, 9*(1), 47–69. https://doi.org/10.1108/jhlscm-12-2017-0073.

Ustundag, A., & Cervikcan, E. (2018). *Industry 4.0: Managing the digital transformation.* London: Springer.

Walker, S. (2013). *Handbook of design for sustainability*. London: Bloomsbury.

Wang, S., Wan, J., Li, D., & Zhang, C. (2016). Implementing smart factory of Industrie 4.0: An outlook. *International Journal of Distributed Sensor Networks.* Article ID 3159805, 10. http://dx.doi.org/10.1155/2016/3159805.

Wong, D., & Ngai, E. (2019). Critical review of supply chain innovation research (1999–2016). *Industrial Marketing Management, 82*(2019), 158–187.

Wu, L., Yue, X., Jin, A., & Yen. D. (2016). Smart supply chain management: A review and implications for further research. *International Journal of Logistics Management, 27*(2), 395–417. https://doi.org/10.1108/IJLM-02-2014-0035.

Yosofi, M., Kerbrat, O., & Mognol, P. (2019). Additive manufacturing processes from an environmental point of view: A new methodology for combining technical, economic, and environmental predictive models. *The International Journal of Advanced Manufacturing Technology, 102,* 4073–4085. https://doi.org/10.1007/s00170-019-03446-2.

Zolfagharian, M., Walrave, B., Raven, R., Georges, A., & Romme, L. (2019). Studying transitions: Past, present, and future. *Research Policy, 48*, 103788.

Cold Chain

Cold Chain and Its Application

Dheeraj Kumar, Ravi Kant Singh, and Apurba Layek

Abstract In the global age, customer satisfaction, the safety of food and health products is at the topmost priority. The cold chain technology seems to be an essential aspect of preserving food products, perishables, and delivering them in the right conditions to the open market. Any changes in the supply chain regarding time, temperature, distance, environmental conditions result in the decline of the net present value of the products. This chapter deals with all the factors responsible for the quality assurance of the cold chain products. It gives a complete idea about the mechanism, processes, and monitoring functions, operational conditions for the cold chain logistics. This chapter provides insights on the temperature standards, and organizations set up the regulations for the cold chains. Future aspects related to this technology to produce, store, and handling all the produced goods have been discussed in this article. The primary focus is given to the temperature-sensitive pharmaceutical products, vaccines, distribution practices, quality management aspects for the cold chain technology.

Keywords Cold chain · Perishables · Temperature standards · Transport integrity · Monitoring · 3Pe's

1 Introduction

In the global world, the relative distance between the different regions of the world has become much smaller. However, in true response, the physical response of the regions is still a genuinely challenging task. As much as the physical separation of the produced goods become more, it can damage the operations involved in multifaceted transport. In some cases, damaging product, goods can occur due to shocks or extreme temperature variation. For perishable goods, particularly food products, its quality degrades with time. The reason behind this concept is that these products maintain chemical reactions, which mitigates as the temperature of the environment

D. Kumar (✉) · R. K. Singh · A. Layek
Department of Mechanical Engineering, National Institute of Technology Durgapur, Durgapur, West Bengal 713209, India
e-mail: dknitdgp1@gmail.com

K. Kumar and J. P Davim (eds.), *Supply Chain Intelligence*, Management and Industrial Engineering, https://doi.org/10.1007/978-3-030-46425-7_4

lowers (Brzozowska et al. 2016). Every delay response in the shipment can have a negative impact if the cargo is a perishable product. Hence it takes time and effective coordination to effectively move the shipment from one market place to another. The cold chain ensures the cargo is not damaged or compromised in the entire process of shipping, the business of pharmaceuticals, medicals, and the food industry. Hence, the industries of concern are moving its mediation towards cold chain or relying on cold chain technology (Bogataj 2005).

The cold chain includes all the transportation of temperature-sensitive products along a supply chain. The methods of the cold chain for packaging the products are thermal and refrigerated techniques. Logistic planning protects the products via the integrity of the shipment. Thus, the cold chain itself is a science, process, and technology. Since it requires the understanding of processes related to chemical and biological, it is a science. It is often called a technology since it requires the knowledge of those techniques to ensure the temperature conditions of the supply chain. It is a process since it includes many tasks to be performed like, preparation, storing the products, transportation, and monitoring the products which are sensitive towards temperature (Brison and LeTallec 2017).

From the perspective view of geography, cold chain technology has the following aspects:

- **Global** permitting the temperature-sensitive products of the agricultural functions to distant products. It enables distributing the pharmaceuticals products, biological, and vaccines from a single large facility to the open market around the global world.
- **Regional** it can involve the facility of cold storage services as regional grocery market or laboratories where the exchange of components related to temperature-sensitive products is taking place.
- **Local** under this category grocery stores, restaurants, final consumer of perishable goods comes to whom the timely distribution of goods can be made successfully.

2 Emergence of the Cold Chain

Since 1950, the providers, founders of third-party logistics, are giving their lot of effort in finding the new methods and emerging technology for the global cold chain commodities. In previous days of the fifties, processes of the cold chain were managed mostly in the houses of manufacturers or the distributors. In the United States nations, the stability of the cold chain was measured by the Food and Drug Administration restrictions and accountability. These companies were relying on couriers rather than complete overhauling of their supply chain facilities (Faisal et al. 2007).

The reliance on the cold chain continued to gain more importance. Now onwards, pharmaceutical industries are focused on testing, production, and movement of drugs with full control and uncompromised transfer of shipping elements. The present scenario shows that the significant portion of the pharmaceutical products of cold chains is experimentally sound and in the developmental phase. About 10% of medical drugs are temperature-sensitive. If the shipment of such products falls under the unanticipated exposure to the varied temperature levels, it may come into the risk failures, resulting in ineffective or harmful to the patients (Bogataj 2005).

Among the supply chains, the cold chain logistics favor their sensation towards the temperature integrity. It involves a higher level of control of the processes included in the cold chain system. It may also provoke the providers of third-party logistics to acquire elements of the supply chain. Day by day the developing countries are paying their attention to the economy towards foods and productions, their needs to keep the product as much as fresh for the extended periods. The cold chain is also focused on the health-related issues of the public. It is because proper transportation will be able to reduce the chances, likeliness of microbial, bacterial growth, fungal contamination of the shipping items. It also handles the transport of medical goods over long distances (Brison and LeTallec 2017).

3 Elements of the Cold Chain

A cold chain technology is a temperature-controlled supply chain system. It involves the product items keeping cold in the whole transaction from point to manufacture to the point of use (Fig. 1).

The elements of the cold chain are considered as the close interaction between these three technologies.

Product

A product has the physical attributes which require to be fulfilled the primary conditions of specific temperature and humidity. The conditions are dictated by transport phenomena, which must not be physically attributed to an unacceptable extent. These attributes are related to the questions of how a product can be fragile and perishable,

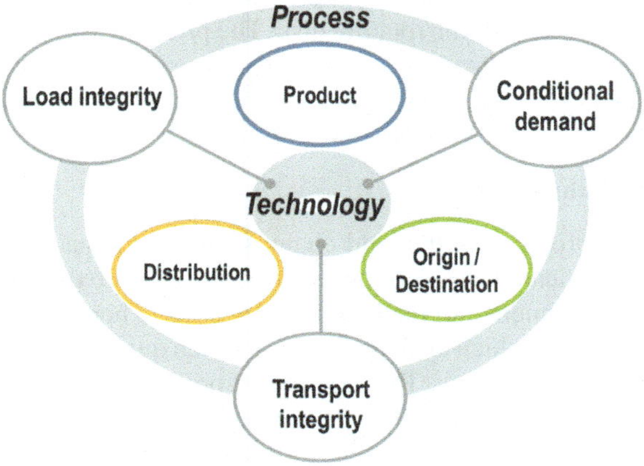

Fig. 1 Elements of the cold chain

how it is handled in the entire cold chain process. Otherwise, the product can come in the category of losing its commercialized value (Joshi et al. 2009).

Origin or destination
It is concerned about the respective locations where the products sensitive towards temperature can be produced or consumed. It deals with the significant constraints and underlines the difficulties coming in the pathway of making products in a market. Because of the cold chain advancement in logistics made it possible to use the increasing strategic sourcing.

Distribution
It includes all those methods and infrastructures needed in the transportation of product goods in a temperature-controlled environment.

4 The Cold Chain Technology

There are few technologies available for the interaction purpose in a sequential manner to support the cold chain technology. It can be figured out in the following way:

- **Monitoring**

This technology governs the connected systems and devices, which monitors the overall condition of the cold chain system. These monitoring conditions are humidity and temperature. Most of the stages where these conditions should be involved are reefer and at the warehouses. The benefits of these technologies are that they provide integrity account for the chain, which helps to find out the potential weakness of the cold chain system. For example, purpose, ISO 10368 standard in 1992 was established to give the guidelines on how the reefer's temperature should be monitored. Previously, for this task to be executed, Partlow recorders were used to measure temperature during transportation. It was recording the fluctuations of the temperature of the rotating disk. The problem associated with this entire process is to retrieve the physical records. Therefore, these outdated devices have been replaced with electronic devices that can be controlled remotely also and can communicate the events and logs (Badia-Melis et al. 2018).

- **Fabrication**

The products of the cold chain, such as pharmaceuticals and foods, are processed or fabricated in a specialized manner, which requires some particular type of types of equipment and method of application. There are some examples of such kind of machines which are blast freezers which quite capable of freezing as soon as possible in the earlier way. This equipment prevents the formation of damaged ice crystals. Once the product is being prepared and finally ready for the shipment, there are so many techniques available for the packing purpose. So, it should be maintained at the desired temperature level for that particular product. Some of the examples are crates, perforated boxes that keep the desired temperature integrity, and protects from the damages also. Vacuum packaging is a renowned technique for the packaging of meats and the extension of its shelf life (Adekomaya et al. 2016).

- **Storage**

The products of the cold chains are rarely available for direct human consumption as finalized goods, like other goods. Therefore, it must be stored in a cold storage facility. The cold chain products are being stored in the refrigerated warehouses unless and until an order has been filled. Specialized distribution centers are being designed in such a way that it may provide all those essential needs to store the grocery goods before coming to the store.

Now, elaborating the distribution section of industrial organizations having a large volume of grocery sections (Fig. 2).

The entire system of food distribution must have to fulfill the demand in a timely fashion, mainly for the distribution of perishable food products. Figure 2 has a 1.1 million square foot Regina warehouse, which has the following subsections as a chamber such as a grocery, produce, dairy and meat, frozen sections. Each of these chambers a particular operating temperature zone (Likar and Jevšnik 2006).

- **Terminal**

For the cold chain logistic purpose ports, airports are the transport terminals used for the shipment of chain colds goods in a growing manner day by day. A container port terminal has that much of substantial available space so that it can provide convenient store refrigerant stores. They also have facilities like dock refrigerated warehouses

Fig. 2 The grocery section of the food distribution center (*Photo* Dr. Jean Paul Rodrigues, 2013)

for storage purposes. Terminals provide their yard space to store the reefers. A figure has been shown of Porte Oceane terminals, which have been established in 2008 have a capacity of 438 plugs. Reefer racks supply communal power supply, and each of the individual reefers can be monitored through the central platform. The overall system requires continuous maintenance of the equipment of the reefers refrigerating. In this process, the temperature of the cargo is tried to keep in the specific range if it failed to keep the balance the operator is subjected to liability (Heap 2006) (Figs. 3 and 4).

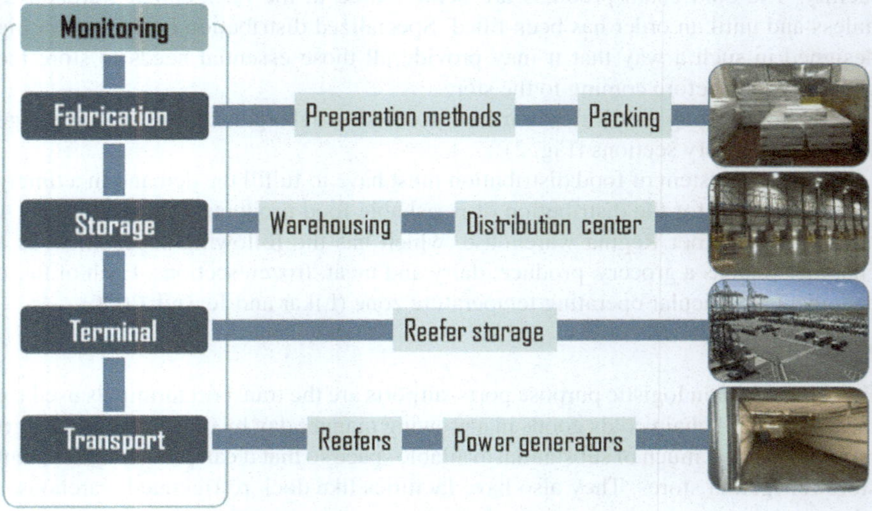

Fig. 3 The cold chain technology

Fig. 4 Reefer rack storage, Porte Oceane Terminal, Le Havre (Dr. Jean Paul Rodrigues, 2010)

Fig. 5 Stacked reefer storage, Maher terminal, New York (*Photo* Dr. Jean Paul Rodrigues, 2010)

- **Transport**

The transport technology is improvised in these days for the active transportation of the cold chain goods. The most common examples of the reefer's vehicles are trucks, unit load devices, containers. Most of the time, maritime containers are being used for this purpose. A photograph is shown of stacked reefer storage at a dedicated area Maher terminal New York, where electric plugs are available. A forklift designs a small platform. So that yard worker might be able to unplug or plug stacked reefer and also can monitor the temperature readings (Fig. 5).

5 Operational Conditions of the Cold Chain Logistics

There are three basic operational conditional processes of a cold chain logistic:

- **Conditional demand**

The demand for any product in a market mainly depends on the qualitative attributes. It has been found that each product has its level of perishability. From an example, it can be easily understood, the value of the vaccine drops to zero value if the shipment occurs is slightly damaged, since in this condition, this product cannot be used. Now another example is in the case of fruits and vegetables, and their value starts declining when the transit place in not maintaining that much quality; as a result, it starts spoiling. Because these products have a limited shelf life (Coulomb 2008).

It has been listed out the shelf life of some selected perishable food products in Table 1.

Table 1 Shelf life of some selected perishable food products

Product	Shelf life (days)	Optimum temperature (°C)
Apple	90–240	0
Bananas	7–28	13.5
Bell peppers	21–35	7
Cabbage	14–20	1
Eggs	180	1.1
Onions	30–180	1
Lettuce	12–14	0.6
Fresh meat (beef, lamb, pork, poultry)	14–65	−2
Oranges	21–90	7
Pears	120–180	−0.6
Potatoes	30–50	10
Seafood (shrimp, lobster, crab)	120–360	−17.8
Strawberries	5–10	0.6
Tomatoes	7–14	12

Source Adapted from APL

The food products start deteriorating, so it is mandatory to ensure that cargo has been kept in the environment where the storage conditions are at the *optimum level*. Therefore, the shelf life of food products can be maximized. Table 1 depicts optimum operating temperature, and the shelf life is dominating. Multiple cold storage sections have much variety for the temperature distribution (Fig. 6).

- **Load integrity**

This concept relates to the load conditions of any transportation that the product must fulfill its value at the timing of loading. It includes the overall concept of packaging and the environmental conditions inside the packaging before transport. A reefer is a controlling unit of the temperature settings; also, it has all sets of products that are sensitive towards temperature range. Reefer is used for ensuring the load integrity of the transport (Malan et al. 2004).

- **Transport integrity**

Temperature controlled environment can be ensured with the help of a series of tasks and safeguards. During transportation, breach integrity might have occurred at the distribution centers and terminals, which is involved in the transport chain. The temperature of the shipping component is being tracked so that if any deviations occur, it can be mediated immediately. All the stringent required in the transport integrity has been designed to support the cold chain logistics with some specialized terminal facilities.

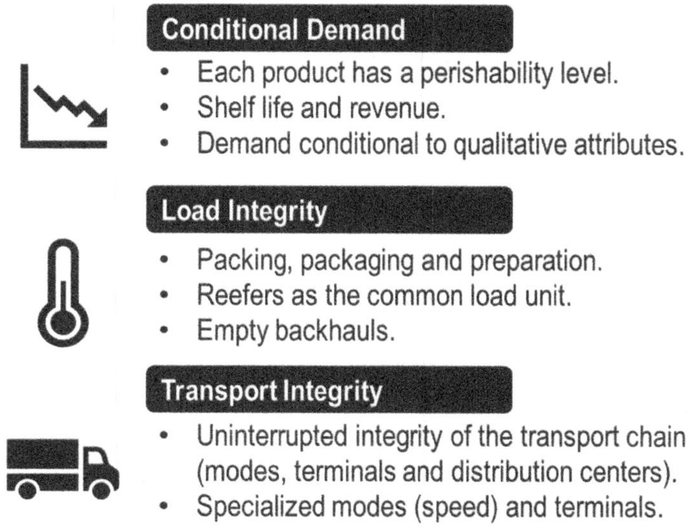

Fig. 6 Operational conditions of the cold chain logistics

6 Sustaining of Temperature Integrity Along a Cold Chain

A chain can be durable as well as weak at its links. This statement is about the relevance of any particular conditions of temperature related to products. A cold chain tries to maintain the temperature of the products within a specific range of 2–8 °C is the typical range. Many foods and pharmaceutical products are like; if the specific temperature range is not maintained, it can damage. Some essential additional critical factors like; packaging properly, temperature, monitoring of all aspects are governed by supply chain integrity. This working concept helps in identifying the location where the breach is taking the place of integrity. Sometimes, it is called the identification of liability. The crucial element of the cold chain, which offers a controlled transport and storage unit, is called reefers. However, it is too large for the many cold chain shipments such as pharmaceuticals (Bo and Danyu 2009) (Fig. 7).

In Fig. 7, it has been shown that the cold chain is maintained along with the carious transport activities having two potential breaches integrity. First, the reefer is being not connected with the power source at the time of transshipment, or it may have an open door for an extended period. Second, it may happen that the product is being kept for in the refrigerated warehouse at a lower temperature than the specified temperature of the product. Since the breaches are not so easy to detect, that is why the challenge remains as it is in which the products are exposed, and some temporary lapses occur in the integrity of the cold chain. As a result, it is observed that a decline in the product's shelf life. Due to the growth of shipments handled by temperature-controlled devices, attention must be given towards identifying the locations, types of equipment, and the circumstances in which a breach in the integrity can take place:

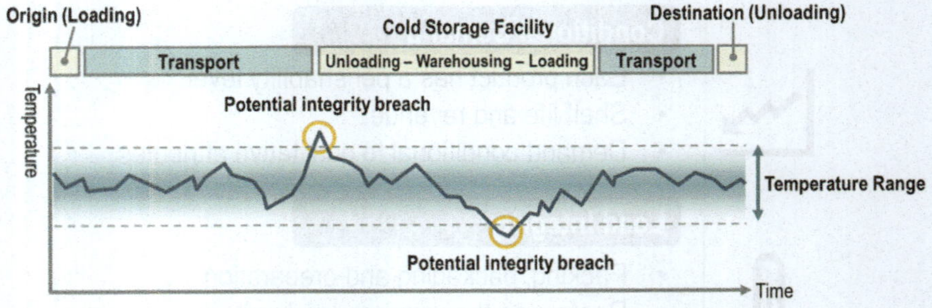

Fig. 7 Maintaining temperature integrity along a cold chain

- **Transportation issues**

During the transportation, cold chain compromises with the malfunctions of the refrigeration in a couple of hours, depending on the ambient temperature. However, the equipment of the refrigeration is made to maintain a specific range of temperatures suitable for cold chains. If any excessive loading occurs to be cooled, it may bring stress to the refrigerant equipment. Due to wear and tear or defective equipment are the responsible factors for the improper environment of cold storage. It is sometimes called poor air circulation or defective insulation of doors. Some more practical reasons may be the cause for the failure of the inadequate functioning of refrigeration; they are (1) to save fuel cost most of the time drivers shut down the refrigeration unit. (2) At the time of delivering the goods, doors are opened for a prolonged period (Saif and Elhedhli 2016).

- **Transshipments and warehousing issues**

Most of the practical scenario is like that in which the cold chain has to be compromised with the situations, like loading, unloading, and warehousing. For instance, products are left on the duct in the loading condition for an extended period, or the refrigeration unit is switched off at the time of transshipment. Some warehouses have poor control and maintenance for temperature control.

7 Temperature Standards for the Cold Chain

The cold chain technology sets up standard ranges of the temperature depending upon the goods to be transported. There are so many known factors which come to the category of damaged food products of fruits or vegetables. Their categories are changes in color, degradation in the texture, rots and molds development, bruising, softening, all of these are the lists through which it comes in the category of degraded food and its degraded market value (Fig. 8).

Five standard temperature settings have categorized which are selected for specific product type:

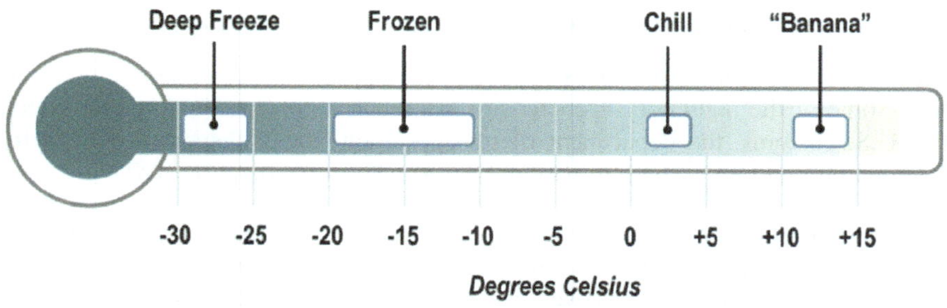

Fig. 8 Temperature standards for the cold chain

- **Deep freeze** (−25 to 30 °C) this temperature range is the coldest temperature range, which has to be achieved by the conventional type refrigerator. This type of refrigerator is mainly for the transportation of ice-creams and seafood.
- **Frozen** (−10 to −20 °C) this temperature range is for products like; frozen meat, beef, poultry, pork, frozen bakery products.
- **Chill** (2–4 °C) this temperature range is the standard range for the refrigerators and commonly used to store food products like; vegetables, fruits, fresh meat.
- **Pharmaceuticals** (2–8 °C) vaccines are transported in this temperature range. However, the products of pharmaceuticals are being transported in the ISO standards of reefers, but for small packages in the refrigerator are transported in vans.
- **Banana** (12–14 °C) this temperature range is the most common range used for the shipping purpose of oranges, pineapples, and vegetables like potatoes.

The above-discussed temperature range is mostly used for transportation. These are quite easy to maintain and monitor. However, reefers can also adapt this range for the maintenance of a specific temperature range of any products to be transported (James and James 2010).

8 Who Is Writing the Rules? Who Is Enforcing Them?

The nature of the cold chain is globally known, although there are some differences in the controlling level of the different parts all over the world. All the controlling authorities have a common goal to give the complete assurance of the stability and quality of the temperature maintenance in the vaccines, pharmaceuticals, and other products of health-related. These authorities have to pay attention to the handling and distribution related issues of the food products. Some standard organizations are being governed to maintain the standard of products, to make the rules and regulations (Bishara 2006).

- **Regulatory agencies, organizations: the guardian of cold chain**

Some administrations govern the standard of quality control for shipping food products. Some of the regulatory agencies are listed below for jurisdiction and orders. The U.S. customs, the department of transportation, the Transportation Security Administration (TSA), The International Air Transport Association (IATA).

US Pharmacopoeia (USA) This agency has the authority to make the rules and regulations for the packaging, manufacturing, and distribution of pharmaceuticals related products. Once it is approved, this becomes enforceable in law and nature. It creates a framework for FDA validation.

Food and Drug Administration (FDA USA) This agency is fully responsible for the control and validating the compliances across the cold chain technology.

Medical and Healthcare Products Regulatory Agency (MHRA Europe) In Europe, this organization has the responsibility for monitoring and enforcement of regulations for medicines.

The World Health Organization (WHO Global) For cold chain packaging, this worldwide organization provides global leadership for the packaging, storage, and handling.

9 Connecting the Dots

Let us discuss a scenario of the pharmaceutical supply chain. In this chain system, the global manufacturing, distribution, and storage are the sub-set linkages of it. This cold chain includes many functions linked with the different potential configurations; for example, outsource partners perform the function of packaging and distribution. In detail, the distribution subdivision has many variables, such as; some product has to be shipped to big pharmacy retailers from the central manufacturer's section. Most of the time, the products are provided through the global networks of distributors and wholesalers (Reed 2005) (Fig. 9).

10 General Queries on the Development of Cold Chain

- **How can the cold chain reduce food loss?**

The cold chain technology reduces the losses of the food products by ensuring an avenue for perishable products to the consumption center. Primarily, the reason for the food losses is only due to unpreserved food coming into contact before consumption. Cold chain technology firstly initiates its steps towards extending the holding life

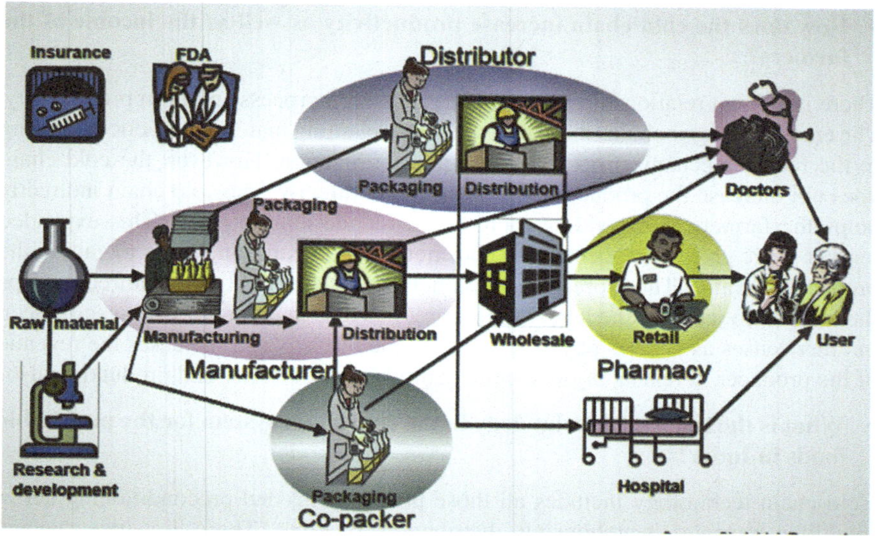

Fig. 9 Detailed view of dots of pharmaceuticals and life science cold chain

of the food product and; secondly, it provides a market to transfer the food product within their life span. In normal conditions, the life span of the perishable fruits and vegetables is limited. However, unless and until the selling period is within its natural time period, this technology mitigates the food losses in the case, when the markets are situated over a long distance.

Cold chain technology does not preserve food products but tries to extend the shelf life of the food products for a predetermined period. It also provides post-harvesting food handling to the organizations. This technology ensures us the damages or stress occurs due to transportation or road conditions (Joshi et al. 2012).

- **How does the cold chain system ensure the safety and quality of the products at the desired level?**

This technology provides an organized handling technique to a structure and also brings compliance with the standards of food safety. The environment provided with this technology reduces the chances of being exposed to the external microbial load with the packaging. This chain ensures safe handling due to grading size. In this technology, the application of the dock levelers helps to achieve good practice ability of the palletized or unit loads in cold chain applications. This technology maintains a healthy environment by providing the correct temperature so that the decomposition of bacteria influences and other activities of enzymes. The cold chain maintains living conditions all about humidity, replenish the air, it segregates to avoid the tainting in between living tissues and keeps the temperature in the optimum condition.

- **How does the cold chain increase productivity as well as the income of the farmers?**

There is a direct relationship between the production process and farm productivity. The entire process includes the aspect of quality planting materials, practices running in the farms, mechanization, irrigation, and fertilization. However, the cold chain does not increase the productivity of the farmers directly. The cold chain indirectly helps the farmers by providing more space for the market place. This expanded market space gives more benefits as enormous market capturing leads to gain in the productivity value. Therefore, it can be concluded that the cold chain indirectly helps the farmers to increase the level of farm productivity. In other words, it can be seen in other senses as it gives physical access to the markets and increases the revenue of the producer. It results in substantial gainful in productivity and production also.

- **What is the renowned technology in the cold chain system for the perishable foods in India?**

Cold chain technology includes all those primarily needed preconditioning activities like; transport, warehousing, distribution, retailing. These activities manage the life cycle of product, packaging, labeling, storing, and distributing, monitoring, atmosphere, refrigeration, food safety, and trade process. In cold chain technology, specific lists are being given below to improve the operational concerns and helps in managing the things. They are as follows:

Warehouse management system, system tracing food, thermochromatic inkpots, equipment of photo-catalytic, packaging like MAP, organic, dynamic racking, CA tents, PLC, various hybrid application in the sourcing of energy system (Gogou et al. 2015).

- **Cold chain technology comparison with other Asian countries**

Based on the strategic national agenda, technology adoption is linked with the business and also linked with the support of the government. It is to be expected that most of the technologies concerned with the government sector promote the adaptation of technology. In the current situation, the focus has been shifted towards providing nutritional security and food safety (Shabani et al. 2012).

In a survey of Global Cold chain Alliance (GCCA) and international association in 2014, it has been recorded different capacities of the refrigerated warehouse in different countries are:

(a) India 131 mill m^3 (b) China 76 mill m^3 (c) Indonesia 12 mill m^3 (d) Bangladesh 0.49 mill m^3 (e) Nepal 0.26 mill m^3

For India, Bangladesh, Pakistan, Nepal, cold chains are mainly concentrated on potatoes. Thailand and China are reported for a large number of pack-houses.

11 Barriers to Overcome: The Principles of 3Pe's

After deciding the objectives of the future and performance targets, it becomes necessary to understand all those factors which, can be an obstacle for the cold chain technology. A practical and flexible model is used to analyze the supply chain and to create a roadmap across the various geographic environments or industries. Figure 10 illustrates how 3Pe's achieve their goal (Reed 2005).

- **Policy: ensures alignment across the cold chain**
  For an organization, it becomes necessary for all human resources working for the execution of that particular task. So, it can be concluded that all the key players try to adapt to the best possible practice in their functional area (Aramyan et al. 2007). They are as follows:
  - **Manufacturers** make a sustainable model which has the property of the sharing the information of the products, total life cycle management. The manufacturer, as an owner of the brand, ultimately, they have the responsibility for ensuring the other cold chain players working or at any point of consumption, that all are correctly doing their job and giving productivity.
  - **Wholesalers** In the cold chain, they do protection of the products from the manufacturer to storekeeping the temperature-controlled and monitoring the distribution process.

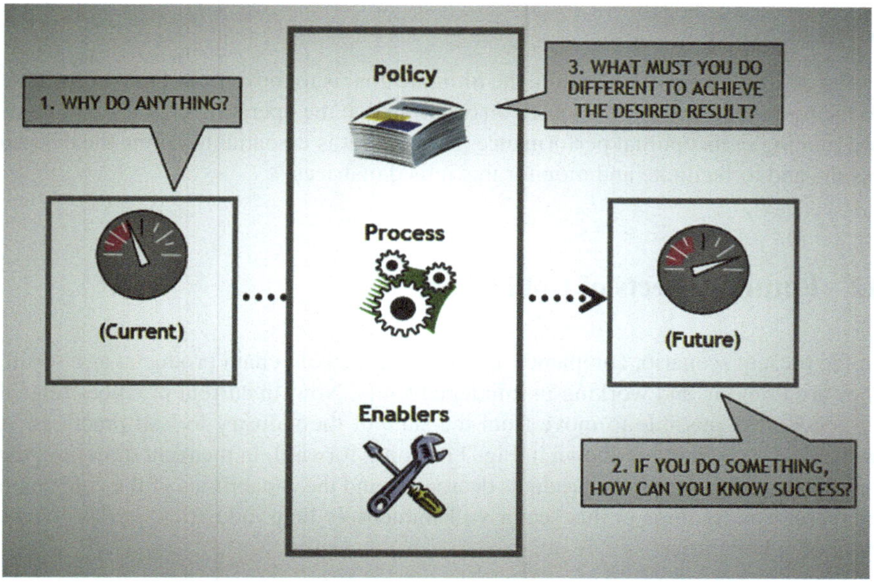

Fig. 10 Improved supply chain: 3Pe's act as a compass

   - **Healthcare providers**, they fulfill all the requirements for storage and handling purpose, ensuring that among all the patients and caretaker the communication should be in a proper way so that no communication gap occur regarding any issues.
   - **Pharmacies and insurance providers** this policy integrates the requirement of cold chain in dispensaries and makes patients automated, which able to follow some selected products. The insurance providers ensure that cold chains are adopting the best practices across the global supply chain with the creation of incentives and audits.

- **Processes: all about managing the links!**

In a cold chain process, a good understanding of links attached and activities at each node from the start of the discovery to the consumption of the product enables all the players collaboratively working under the organization and ensuring the optimal delivery model.

- **Enablers: making the business case**

The best combination of improved policy, process, performance, and enablers of the technology makes it possible for all players in the worst critical condition to achieve the incremental achievement in productivity. The transition period from the better to best one, adopted by the players of the cold chain, ensures that the products of the cold chain are distributed with the same level of care of the manufacturing process to ensure the quality and consistency (Faisal et al. 2007).

- **Performance**

In the entire process of operation, the ultimate goal is to move from the present state to the best practice state. It focuses on ensuring that the operating chain components are running at its optimal performance. However, it is essential to define the desired results and to facilitate and monitor the agreed matrices.

12 Future Aspects of Cold Chain

In the present scenario, companies are aware of the cold chain products; as a result, they are thinking and working in it independently. Now, in current practices related to goods, it is possible to move from the state of the industry to best practices. A well-illustrated figure is shown in Fig. 11. Going forward, in the right direction, the risk factor of damaging the products decreases, and the capabilities of the cold chain increase. The findings in this sector will immensely help industries, professionals, developing countries.

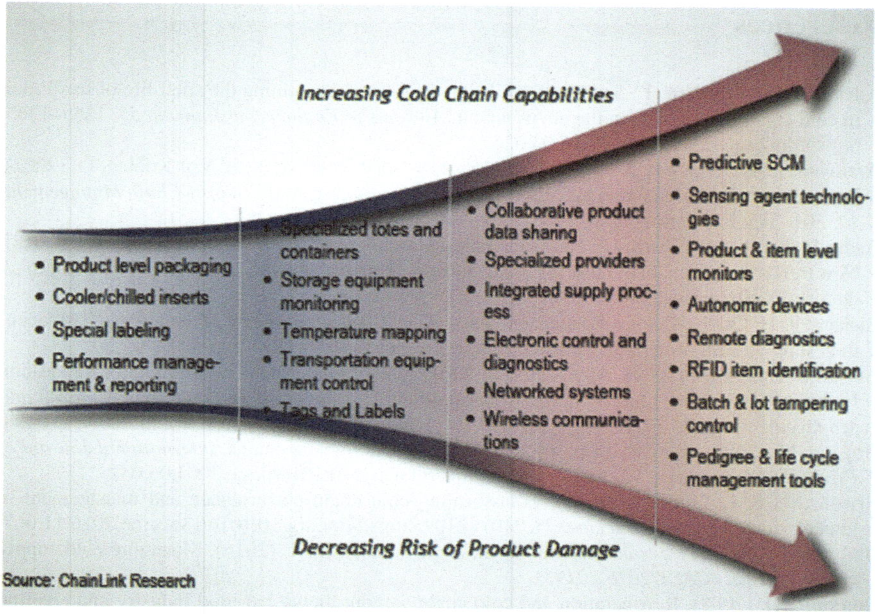

Fig. 11 Future forward progressions of cold chain practices

13 Conclusion

The cold chain technology is the primary domain for the food sectors and pharmaceuticals. The significant findings in cold chain research work will help in imposing the directions and respective domains of many inhibitors to make an efficient cold chain. There are many challenges in the modern era for the cold chain application. This article has outlined many art technologies that individually help in improving the efficiency of the cold chain. This chapter provides a complete insight into the logistic decision models for perishable products in the cold chains. It gives a complete idea about the mechanism, processes, and monitoring functions, operational conditions for the cold chain logistics. This chapter provides insights on the temperature standards, and organizations set up the regulations for the cold chains. The shelf life of the food products can be made longer after getting complete knowledge of the cold chain from this chapter, which enables readers to maintain a packaging environment suitable for transportation without compromising with the quality standard. The entire discussion develops a comprehensive and extensive idea in cold chains, serves as a valuable tool to researchers and academicians.

References

Adekomaya, O., Jamiru, T., Sadiku, R., & Huan, Z. (2016). Sustaining the shelf life of fresh food in cold chain—A burden on the environment. *Alexandria Engineering Journal, 55,* 1359–1365. https://doi.org/10.1016/j.aej.2016.03.024.

Aramyan, L. H., Lansink, A. G. J. M. O., Van Der Vorst, J. G. A. J., & Van Kooten, O. (2007). Performance measurement in agri-food supply chains: A case study. *Supply Chain Management, 12,* 304–315. https://doi.org/10.1108/13598540710759826.

Badia-Melis, R., Mc Carthy, U., Ruiz-Garcia, L., Garcia-Hierro, J., & Robla Villalba, J. I. (2018). New trends in cold chain monitoring applications—A review. *Food Control, 86,* 170–182. https://doi.org/10.1016/j.foodcont.2017.11.022.

Bishara, R. H. (2006). Cold chain management—An essential component of the global pharmaceutical supply chain. *American Pharmaceutical Review, 9,* 105–109.

Bo, Y., & Danyu, L. (2009). Application of RFID in cold chain temperature monitoring system. In *2009 Second ISECS International Colloquium on Computing, Communication, Control, and Management CCCM 2009* (Vol. 2, pp. 258–261). https://doi.org/10.1109/CCCM.2009.5270408.

Bogataj, M. (2005). Stability of perishable goods in cold logistic chains. *International Journal of Production Economics, 93–94,* 345–356. https://doi.org/10.1016/j.ijpe.2004.06.032.

Brison, M., & LeTallec, Y. (2017). Transforming cold chain performance and management in lower-income countries. *Vaccine, 35,* 2107–2109. https://doi.org/10.1016/j.vaccine.2016.11.067.

Brzozowska, A., Brzeszczak, A., Imiołczyk, J., & Szymczyk, K. (2016). Managing cold supply chain. In *IEEE ICALT'2016 2019.*

Coulomb, D. (2008). Refrigeration and cold chain serving the global food industry and creating a better future: Two key IIR challenges for improved health and environment. *Trends in Food Science & Technology, 19,* 413–417. https://doi.org/10.1016/j.tifs.2008.03.006.

Faisal, M. N., Banwet, D. K., & Shankar, R. (2007). Information risks management in supply chains: An assessment and mitigation framework. *Journal of Enterprise Information Management, 20,* 677–699. https://doi.org/10.1108/17410390710830727.

Gogou, E., Katsaros, G., Derens, E., Alvarez, G., & Taoukis, P. S. (2015). Cold chain database development and application as a tool for the cold chain management and food quality evaluation. *International Journal of Refrigeration, 52,* 109–121. https://doi.org/10.1016/j.ijrefrig.2015.01.019.

Heap, R. D. (2006). Cold chain performance issues now and in the future. *Innovative Equipment and Systems for Comfort & Food Preservation,* 1–13.

James, S. J., & James, C. (2010). The food cold-chain and climate change. *Food Research International, 43,* 1944–1956. https://doi.org/10.1016/j.foodres.2010.02.001.

Joshi, R., Banwet, D. K., & Shankar, R. (2009). Indian cold chain: Modeling the inhibitors. *British Food Journal, 111,* 1260–1283. https://doi.org/10.1108/00070700911001077.

Joshi, R., Banwet, D. K., Shankar, R., & Gandhi, J. (2012). Performance improvement of cold chain in an emerging economy. *Production Planning & Control, 23,* 817–836. https://doi.org/10.1080/09537287.2011.642187.

Likar, K., & Jevšnik, M. (2006). Cold chain maintaining in food trade. *Food Control, 17,* 108–113. https://doi.org/10.1016/j.foodcont.2004.09.009.

Malan, D., Fulford-Jones, T., Welsh, M., & Moulton, S. (2004). *Cold chain management using an ultra low power wireless sensor network* (pp. 21–23).

Reed, C. (2005). Cold chains are hot! Mastering the challenges of temperature-sensitive distribution in supply chains. *ChainLink Research.*

Saif, A., & Elhedhli, S. (2016). Cold supply chain design with environmental considerations: A simulation-optimization approach. *European Journal of Operational Research, 251,* 274–287. https://doi.org/10.1016/j.ejor.2015.10.056.

Shabani, A., Saen, R. F., & Torabipour, S. M. R. (2012). A new benchmarking approach in cold chain. *Applied Mathematical Modelling, 36,* 212–224. https://doi.org/10.1016/j.apm.2011.05.051.

To Analyse the Impact and Benefits of Cold Chain Applications for Frozen Food at High-Temperature Zone: A Case Study of Rajasthan, India

Anju Bharti and Shivangi Sahay

Abstract Cold chain is a series of actions and equipment applied to maintain a product within a specified low-temperature range from harvest or production to consumption. In other words, cold chain is a supply chain where temperature is being controlled. Farmers are putting their efforts in using the green revolution technologies for strengthening food and income security, in addition to generating surplus production. India has tripled the growth of food processing industry, increasing its level of processing from 6 to 20%. But, due to lack of proper transportation and cold chain facility, the edibles were getting wasted. If we are able to save food with negligible expense, we will be able to feed the poor population with low cost strengthening the economy of country. We can minimize the expense of food processing procedure through the development and enhancement of basic facilities for storage of agricultural food produce, transportation and processing. The Indian agricultural sector's growth is quite poor due to the lack of infrastructural amenities. The administration must put an effort to allocate funds for the interest of the nation for setting up of warehouses, contemporary widespread market and agro-logistics hub along with cold chain facilities. There is a requirement not only for the cold chain system but also there is a growing need for modern incubation centres, quality testing laboratories and storage amenities for enhancing the process of manufacturing. To cater the needs of the people, the cold chain system for processed food became highly recommendable. Rajasthan was selected for the study which is known for its varied climate. It has all kinds of produce due to rich quality of soil in most places. In this paper, we will observe the effectiveness of cold chain on edibles to retain its quality and to be available for the ultimate customer.

Keywords Cold chain · Frozen food · Wastage · Economy · Profitability · Implementation

---

A. Bharti (✉)
Department of Management, Maharaja Agrasen Institute of Technology, Rohini, Delhi, India
e-mail: bhartianju4147@gmail.com

S. Sahay
Department of Electronics and Communication Engineering, Indira Gandhi Delhi Technical University for Women, New Delhi, Delhi, India
e-mail: sahayshivangi1408@gmail.com

K. Kumar and J. P Davim (eds.), *Supply Chain Intelligence*, Management and Industrial Engineering, https://doi.org/10.1007/978-3-030-46425-7_5

1 Introduction

India is a major producer of food (fruits, vegetables, wheat, pulse, milk, spices, etc.) in the world after China. This industry has a significant economic component as this sector has been employing nearly 60% of the country's population being the front end of the agriculture sector. It is contributing to around 25% of India's gross domestic product. India has an access to a large natural resource base of 161 million hectares of arable land and 15 million hectares of freshwater reservoirs. It also has the largest livestock population in the globe and diverse agro-climatic conditions. Due to all these resources, India has become a favourable destination for growth in the food industry. The Indian food industry is expected to reach $258 billion from the current level of $181 billion. India is on the threshold of overtaking Japan to become the third-largest country in the world. The predictions state that India will rank among the world's top five economies by GDP in 2025, up from tenth today. In addition, in the next few decades, the country will likely surpass China as the world's most densely inhabited country. As a result, multinational consumer goods companies seeking faster growth must focus on the Indian subcontinent. India has focused on foreign direct investment, so its investment climate is improving. Agricultural sector is a significant economic component, employing nearly 60% of the country's population and contributing to around 25% of India's gross domestic product (Bharti 2016).

The growing urbanization, growing incomes of every member of family and rising aspiration for a superior life, particularly, among the lower economic strata are a few of the factors helping in restructuring the Indian consumer marketplace. Even after the high production of food, the need for food is not being fulfilled (Bharti 2014). In Rajasthan, several crores worth of food goes waste due to lack of cold storage and proper warehouse facilities (Bharti and Goyal 2017).

Indian economy is predominantly an agrarian economy, and agriculture is the backbone of the Indian Economy. Indian agriculture is the largest private sector enterprise in the country of over 110 million farmers. It engages two-third of total personnel, supply to 26% of the GDP and practically one-sixth of the total sell overseas wage.

Factors that are likely to grow in demand for processed and frozen food are as follows:

- There is a drastic change in lifestyles and growth in disposable income. There is also a rise in double income in families and proportion of women in the workforce in the organization.
- To make the processed food more affordable, there is decrease in prices of processed foods, thereby accessing a much larger market.
- A variety of retail formats are being developed, and rapid growth was registered in organized retail (>20% p.a.).
- It is being estimated that investment of about $22 billion will be done in the next 10 years.
- There is a rise in household incomes due to increase in urbanization, changing lifestyles of people. It has led to the rapid growth of the private sector. The

dairy-processing industry will lead to greater demand for value-added, milk-based products, like the processed cheese, table butter and ice cream liked by all.

- Rise in incomes will make fish more affordable for a larger segment of the population.
- Fruit consumption increased at a CAGR of 4.33% from 2007 to 2011, which was the highest among all the food products report.
- The popularity of fruit juice drinks and bottled water has increased as the growth rate of soft drink sales has decelerated due to pesticide contamination issues.
- Coffee consumption had expanded at a rapid rate during the forecasted period—increased at a CAGR of 10.05% for the period spanning from 2007 to 2011 where processed milk food is also being used.

1.1 Objective

1. To understand the impact of technology in improving the effectiveness and benefits of cold chains on edibles at selected outlet.
2. To study the influence of effective applications of cold chain for frozen products on the profitability and improving footfall of selected retail outlets of Rajasthan (India) (Diagram 1).

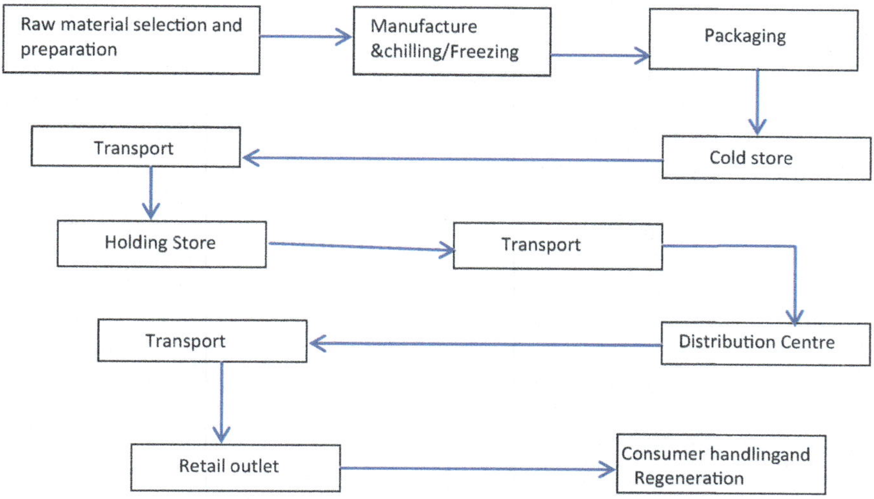

Diagram 1 Flow chart of cold supply chain (*Source* seafood.oregonstate.edu, 1999)

1.2 *Literature Review*

1.2.1 The Supply Chain and the Cold Chain

The supply chain has more than one definition. Kovacs (2004) stated that supply chain is a 'product-based approach' of industrial ecology. It includes the companies that provided the products, raw materials, technology and service. Furthermore, companies which helped to make the operation of the production chains can also be considered as part of the supply chain. Maxwell et al. (2006) argued that both the supply chain and cold chain management could be defined under the 'umbrella' of the sustainable production and consumption. Given above, the supply chain meant that sustainable operation and coordination of the companies were involved in the process of providing the products.

Joshi et al. (2009) stated the two main differences between supply chain and the cold chain were that firstly the cold chain demanded a lot on the operating conditions; secondly, from the production spots to the consuming place, products in the cold chain have the possibility to be spoiled. Salin et al. (2003), meanwhile, explained that the relationship between supply chain and cold chain is that the cold chain can be viewed as facilities and conditions demanded in a supply chain, whereas Maxwell et al. (2006) suggested that the sustainability is important for both of them.

Lisa (2013) discussed that a cold chain for perishable foods is the uninterrupted handling of the product within a low-temperature environment during the post-harvest steps of the value chain.

Packaging
Nielsen (2006) described that there is an influence on the fast-moving consumer goods (FMCG) markets due to increased prevalence of refrigerator ownership which has expanded from 48 to 68% during 1996–2006 period. This increase has given more consumers the ability to buy food products from the cold chain.

The Fresh Food Logistics
According to Joshi et al. (2009), the fresh food logistics referred to a series of logistics processes that are conducted in the cold chain, which includes transportation stage (handling, loading, delivery) and storage stage (warehousing, cooling and packaging). In this chapter, the fresh food logistics means the logistics procedure from the production points to the consumers, which contains both transportation and storage.

Opportunity in India
According to Viswanadham (2005), India still does not have a comprehensive cold chain network. The cold chains which are existing are found to be underutilized. There is a lack of even refrigerated transportation as a system is almost non-existent in India. The farmers have been using mostly open trucks to send their produces to the mandis at a high percentage loss, probably because of the high cost of refrigerated transport. There is a big opportunity in India for infrastructure builders, cold chain

operators, logistics companies and food manufacturers both from inside and outside of India for investment.

The Cost and Effectiveness of Fresh Food Logistics
Liu et al. (1999), Joshi et al. (2009), Boronico et al. (1996), Beardsell et al. (2002), Kelepouris et al. (2007), Henderson (1994), Manos and Manikas (2010) and Manning et al. (2006) conveyed that many perspectives can be adapted to discuss the fresh food logistics; however, the literature study indicates that analysing the problem via the perspective of cost and effectiveness is main concern.

Practical storage times for various foods at a temperature of −18 °C are given in Table 1.

1.3 Research Design

In the present study, exploratory type of research has been adopted in the first step which was followed by the descriptive type of research as very limited study has taken place related to the topic of present research, related to cold chain management specifically in Rajasthan (India) at selected outlets.

The rationale behind using exploratory research design was that not much literature was available on the subject that would have fulfilled the objectives of research. This had necessitated gathering of first-hand information from those in cold chain and frozen food retail business. The chosen research design mainly emphasizes on the discovery of ideas and development of insight into the subject under study.

1.4 Questionnaire Development

Researcher had mostly used a personal approach (through interviews/telephonic interviews) with employees of big and small stores because they had better experience about the subject. Managers of retail stores were interviewed to understand to identify the factor determining the effectiveness of cold storage. The focus of these semi-structured interviews was to understand the goals/drivers to identify the factor determining the effectiveness of cold storage.

1.5 Methodology

The primary data obtained through questionnaires will be analysed by using appropriate statistical tools. The filled-up questionnaires will be coded, and master data sheet will be prepared. The data then will be tabulated and classified on the basis of independent and dependent variables. The summary statistics of the collected data

Table 1 Maximum storage times for frozen foods at −18 °C

Edible products	Practical storage life (in months) at −18 °C
Green vegetables/potato	
Broccoli	18
Green beans	15
Carrots	18
Cauliflower	15
Corn on the cob	12
Peas	18
Potato chips	24
Spinach	18
Raw meat and meat products	
Beef joints, steaks	12
Beef mince	10
Lamb joints, chops	10
Pork joints, chops	6
Sausages	6
Bacon	2–4
Chicken, whole	18
Chicken, portioned	18
Turkey, whole	15
Duck/geese, whole	12
Fish and shellfish	
Oily fish (e.g. herring, salmon, mackerel)	4
White fish (e.g. sole, plaice)	8
Flat fish (e.g. sole, plaice)	10
Prawns, lobster, crab	6
Clams, oysters	4
Other foods	
Ice cream	6

Source seafood.oregonstate.edu, 1999

will be presented in systematic manner. Thereafter, different null hypotheses will be developed to identify the influence of certain independent variables on effectiveness of cold chains for frozen food products. Various hypotheses have been derived and were tested (Bharti 2017). The statistical tests, regression, multicollinearity, variation inflation factor (VIF) analysis will be used for analysis.

1.6 Analysis of Benefits of Cold Chain

This section of the study discusses the benefits of cold chain and its impact on profitability and footfall of the retail store. There are a lot of factors that affect the profitability of retail store, and effective cold chain is one of them. In this section, various aspects of a cold chain have been studied. A cold chain helps in maintaining the quality, texture, odour, taste and moisture of products, and as a result, it saves the store from probable loss. This finally increases the profitability of the store. Similarly, if an effective cold chain is maintained by a store, the customers start believing in the freshness and quality of the frozen products of those stores; hence, the footfall in the store increases.

This section of the study has been divided into two parts (Bharti 2017):

1. **Analysis of various benefits of a cold chain**: With the help of extant literature and informal discussions with the retail store managers and experts, 12 benefits of cold chain were listed. The respondents are asked to respond on a five-point Likert scale (from strongly agree to strongly disagree). Mean and standard deviation have been calculated to analyse the data for the same.
2. **Impact of various benefits of a cold chain on overall profitability of a store and customer footfall in the store**: It has been found and discussed that which variables under a cold chain affect the profit and 'footfall'.

To check the significance of those variables, multiple regression (stepwise) has been applied with dependent variables:

1. **An Effective Cold chain increases the Overall Profits of a retail Store**, and
2. **An Effective Cold chain is Mandatory for good footfall in a retail store.**

2 Methodology for Calculation of Mean

Table 2 shows that a large number of respondents (store managers/assistant managers, etc.) are strongly agree with the statement 46.4% followed by 32.1% respondents who are 'agree' with the statement—'**It maintains the quality of the products**'. 17.9% are neutral, and around 0.9% of the respondents are disagree and 2.7% strongly disagree. This shows that most of the respondents are towards agreement side of the statement. The mean value of the statement is 4.19 which is also above the 'agree' value of the scale (refer Table 2) (Fig. 1).

Table 3 shows that a large number of respondents (store managers/assistant managers, etc.) are strongly agree with the statement 25 and 42.9% respondents who are 'agree' with the statement that '**It is very important for storage of vegetables**'. However, at the same time, a large number of respondents—28.6% are neutral to the statement and 2.7% are disagree and 0.9% strongly disagree. This shows that most

Table 2 It maintains the quality of the products

Response categories	Frequency	Per cent	Cumulative per cent
Strongly agree	52	46.4	46.4
Agree	36	32.1	78.6
Neutral	20	17.9	96.4
Disagree	1	0.9	97.3
Strongly disagree	3	2.7	100.0
Total	112	100.0	
Mean 4.19			

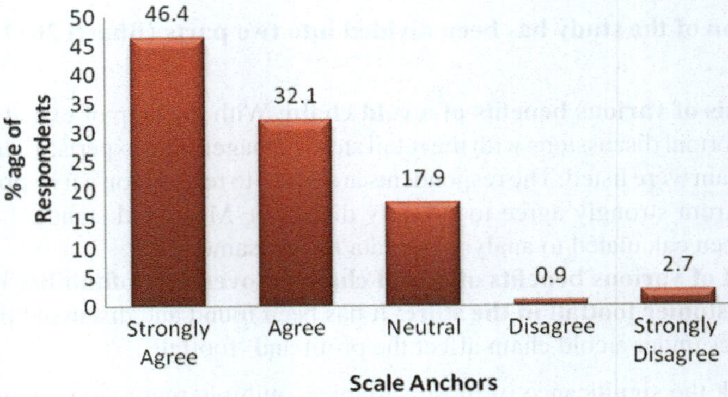

Fig. 1 It maintains the quality of the products

Table 3 It is very important for storage of vegetables

Response categories	Frequency	Per cent	Cumulative per cent
Strongly agree	28	25.0	25.0
Agree	48	42.9	67.9
Neutral	32	28.6	96.4
Disagree	3	2.7	99.1
Strongly disagree	1	0.9	100.0
Total	112	100.0	
Mean 3.88			

of the respondents are towards agreement side of the statement. The mean value of the statement is 3.88 which is slightly below the 'agree' value of the scale; hence, it may be concluded that respondents are more or less agree to this statement (Fig. 2).

Table 4 shows that a large number of respondents agree with the statement (43.8%) and around 25.0% are neutral, followed by 15.2% respondents who 'strongly disagree' with the statement regarding cold chain that '**It helps in restoring the texture**

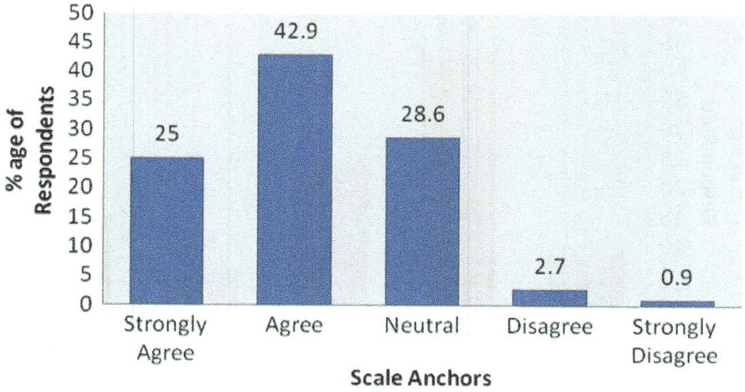

Fig. 2 It is very important for storage of vegetables

Table 4 It helps in restoring the texture of the products

Response categories	Frequency	Per cent	Cumulative per cent
Strongly agree	11	9.8	9.8
Agree	49	43.8	53.6
Neutral	28	25.0	78.6
Disagree	7	6.3	84.8
Strongly disagree	17	15.2	100.0
Total	112	100.0	
Mean 3.27			

of the products'. Only 9.8% strongly agree and 6.3% disagree with this statement. The mean value of the statement is 3.27 which is slightly above the neutral value of the scale. Though this value is moving towards 'agree' value of scale, it will be more justifiable to say that respondents are neutral for statement regarding benefit of cold chain **in restoring the texture of the products** (Fig. 3).

Table 5 shows that a large number of respondents strongly agree with the statement (48.2%) and equal number of respondents agree with the statement that the cold chain **helps in protecting the products from bacteria**. Only 3.6% are neutral, and there is no respondent who disagree or strongly disagree with the statement. This shows that most of the respondents are towards agreement side of the statement. The mean value of the statement is 4.45 which is much above the 'agree' value of the scale and moving towards the strongly agree value (Fig. 4).

Table 6 shows that a large number of respondents are strongly agree with the statement (42.9%) and 55.4% of the respondents agree with the statement that the cold chain **helps in retaining the nutrients of the products**. Only 1.8% are neutral, and there is no respondent who either disagree or strongly disagree. This shows that most of the respondents are towards agreement side of the statement. The mean

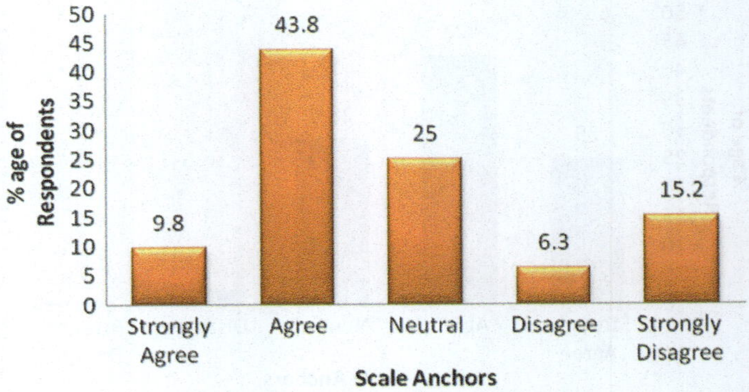

Fig. 3 It helps in restoring the texture of the products

Table 5 It helps in protecting the products from bacteria

Response categories	Frequency	Per cent	Cumulative per cent
Strongly agree	54	48.2	48.2
Agree	54	48.2	96.4
Neutral	4	3.6	100.0
Disagree	0	0	100.0
Strongly disagree	0	0	100.0
Total	112	100.0	
Mean 4.45			

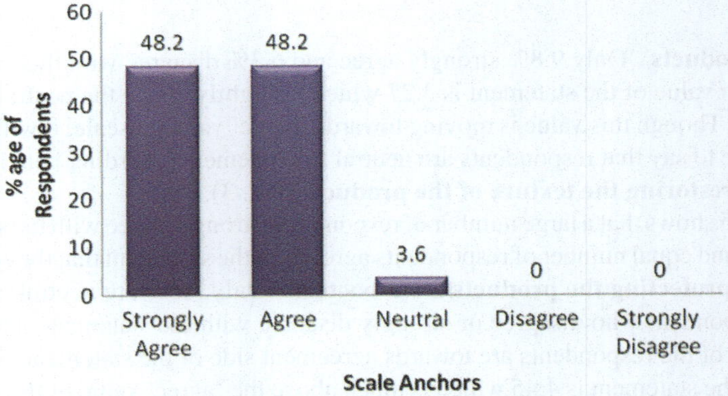

Fig. 4 It helps in protecting the products from bacteria

Table 6 It helps in retaining the nutrients of the products

Response categories	Frequency	Per cent	Cumulative per cent
Strongly agree	48	42.9	42.9
Agree	62	55.4	98.2
Neutral	2	1.8	100.0
Disagree	0	0	100.0
Strongly disagree	0	0	100.0
Total	112	100.0	
Mean 4.41			

value of the statement is 4.41 which is much above the ‘agree’ value of the scale and moving towards the strongly agree value (Fig. 5).

Retaining colour of the product is also an important aspect. Colour is mainly important for fruits and vegetables. Table 7 shows the results for this statement that **cold chain helps in retaining the colour of the products or not**. The results

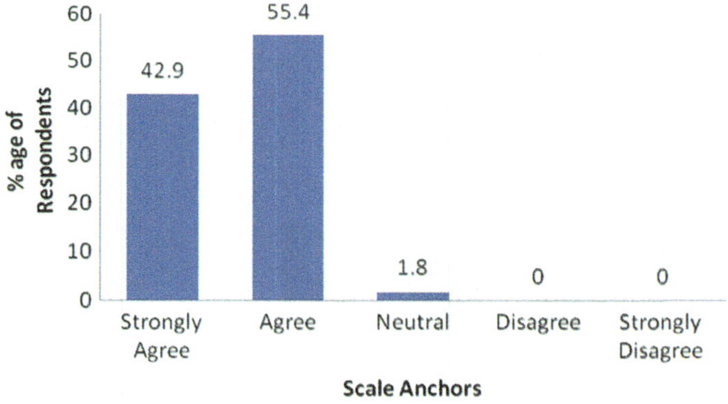

Fig. 5 It helps in retaining the nutrients of the products

Table 7 It helps in retaining the colour of the products

Response categories	Frequency	Per cent	Cumulative per cent
Strongly agree	20	17.9	17.9
Agree	61	54.5	72.3
Neutral	28	25.0	97.3
Disagree	3	2.7	100.0
Strongly disagree	0	0	100.0
Total	112	100.0	
Mean 3.88			

show that a large number of respondents agree to the statement (54.5%); however, only 17.9% are strongly agree. 25% are neutral, and only 2.7% are disagree. There is no respondent who strongly disagree. This shows that most of the respondents are towards agreement side of the statement. The mean value of the statement is 3.88 which is slightly less than the 'agree' value of the scale and very close to the agreement value (Fig. 6).

Cold chains help in retaining the odour of the product As per the responses shown in Table 8, it is found that 31.3% of the respondents strongly agree with this statement, whereas 55.4% are agree. 11.6% are neutral, and only 1 respondent (0.9%) is disagree and strongly disagree. The mean value comes out to be 4.15 which is more than the agree value of the scale. Hence, it may be concluded that the respondents agree with the statement (Fig. 7).

Table 9 shows the responses of store managers and assistant managers regarding the benefits of cold chain i.e.—**cold chain increases the storage and shelf life of the products**. It is found from Table 9 that 37.5% of the respondents are strongly agree with this statement, whereas 58% are agree. 4.5% are neutral and there is no respondent who either disagree or strongly disagree. The mean value comes out to

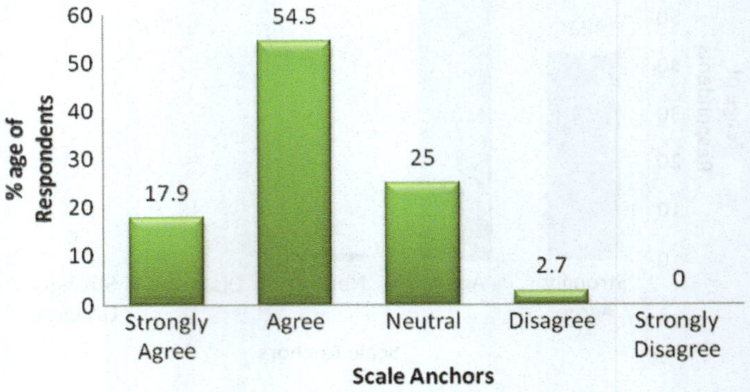

Fig. 6 It helps in retaining the colour of the products

Table 8 It helps in retaining the odour of the products

Response categories	Frequency	Per cent	Cumulative per cent
Strongly agree	35	31.3	31.3
Agree	62	55.4	86.6
Neutral	13	11.6	98.2
Disagree	1	0.9	99.1
Strongly disagree	1	0	100.0
Total	112	100.0	
Mean 4.15			

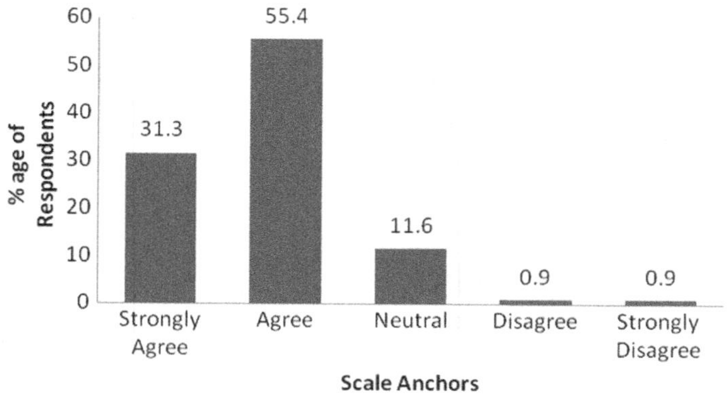

Fig. 7 It helps in retaining the odour of the products

Table 9 It helps in increasing the storage and shelf life of the product

Response categories	Frequency	Per cent	Cumulative per cent
Strongly agree	42	37.5	37.5
Agree	65	58.0	95.5
Neutral	5	4.5	100.0
Disagree	0	0	100.0
Strongly disagree	0	0	100.0
Total	112	100.0	
Mean 4.33			

be 4.33 which is more than the agree value of the scale; hence, it may be concluded that the respondents agree with the statement and intending towards the 'strongly agree' anchor of the scale (Fig. 8).

Table 10 shows that a large number of respondents are strongly agree with the statement (42.9%) and 51.8% of the respondents are agree with the statement that the cold chain **helps in minimizing the spoilage and wastage**. 5.4% are neutral, and there is no respondent who disagree or strongly disagree. This shows that most of the respondents are towards agreement and strongly agreement side of the statement. The mean value of the statement is 4.38 which is much above the 'agree' value of the scale and moving towards the strongly agree value (Fig. 9).

Table 11 shows the responses for the statement that '**An Effective Cold chain increases the footfall in a retail Store**'. This has been taken as a **Dependent Variable**, which depends upon the other benefits provided by the cold chain (this will be later used in stepwise multiple regression analysis). 50.9% respondents are strongly agree with the statement, and 33.9% of the respondents are agree with the statement. 8.9% are neutral, and 6.3% are disagree with the statement. There is no respondent who strongly disagree with the statement. The mean value of the statement is **4.29**.

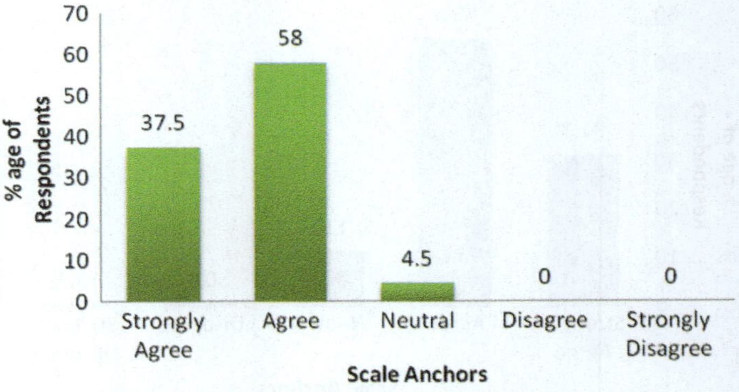

Fig. 8 It helps in increasing the storage and shelf life of the product

Table 10 It helps in minimizing the spoilage and wastage

Response categories	Frequency	Per cent	Cumulative per cent
Strongly agree	48	42.9	42.9
Agree	58	51.8	94.6
Neutral	6	5.4	100.0
Disagree	0	0	100.0
Strongly disagree	0	0	100.0
Total	112	100.0	
Mean 4.38			

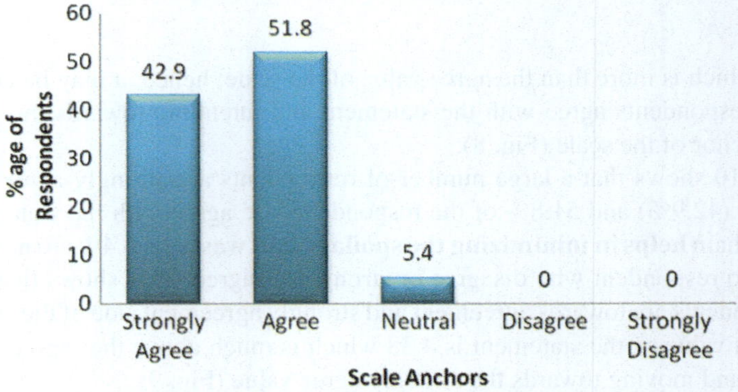

Fig. 9 It helps in minimizing the spoilage and wastage

Table 11 An effective cold chain increases the footfall in a retail store (dependent variable)

Response categories	Frequency	Per cent	Cumulative per cent
Strongly agree	57	50.9	50.9
Agree	38	33.9	84.8
Neutral	10	8.9	93.8
Disagree	7	6.3	100.0
Strongly disagree	0	0	100.0
Total	112	100.0	
Mean 4.29			

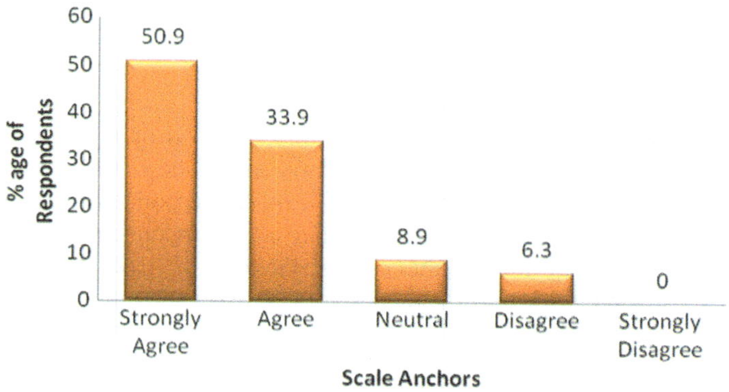

Fig. 10 An effective cold chain increases the footfall in a retail store

It shows that the respondents are agree with the statement and somewhat moving towards strongly agree value of the scale. Hence, the cold chain contributes in the profits of a retail stores (Fig. 10).

Table 12 shows the responses for the statement that **'An Effective Cold chain increases the Overall Profits of a retail Store'**. This has been taken as a **Dependent**

Table 12 An effective cold chain increases the overall profits of a retail store (dependent variable)

Response categories	Frequency	Per cent	Cumulative per cent
Strongly agree	42	37.5	37.5
Agree	64	57.1	94.6
Neutral	6	5.4	100.0
Disagree	0	0	100.0
Strongly disagree	0	0	100.0
Total	112	100.0	
Mean 4.32			

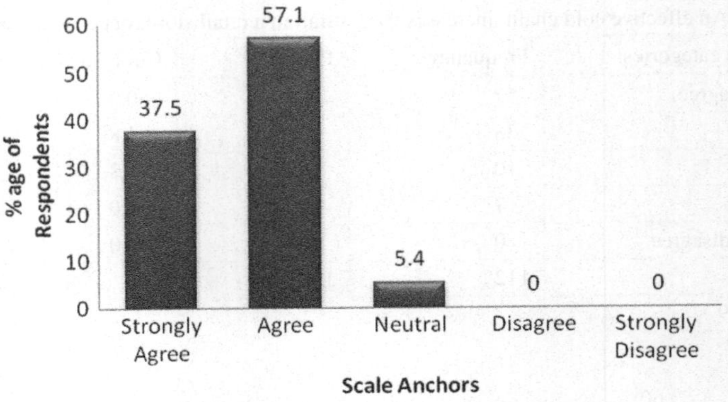

Fig. 11 An effective cold chain increases the overall profits of a retail store

Variable, which depends upon the other benefits provided by the cold chain (this will be later used in stepwise multiple regression analysis). 37.5% respondents are strongly agree with the statement, and 57.1% of the respondents are agree with the statement. Only 5.4% are neutral, and there is no respondent who is disagree or strongly disagree. The mean value of the statement is **4.32**. It shows that the respondents are agree with the statement and somewhat moving towards strongly agree value of the scale. Hence, the cold chain contributes in the profits of a retail stores (Fig. 11).

Table 13 shows the consolidated means and standard deviation for statements related to the benefits of cold chain. Most of the variables have got a mean value of above 4 and which means that cold chain provides these benefits. The most important benefit of cold chain has been identified as the **cold chain helps in retaining the moisture of the products (4.53)** followed closely by **protection from bacteria (4.45)** and protection of **products from humidity (4.44). Retaining the nutrients of the products is yet another benefit of the cold chains (4.41). Cold chain also minimizes the spoilage and wastage (4.38), increases the storage and shelf life of the product (4.33), and helps in retaining the freshness of the products (4.31)**. The mean value for the variable that **cold chains maintain the quality of the product** has also got a good mean value viz. **4.19**.

The least mean value is of the statement that it helps in restoring the texture of the products. The mean value for this statement is 3.27. Two more benefits have been rated fairly by the respondents namely—important for storage of vegetables and retaining the colour of the products with a mean value of 3.88 each.

The dependent variables, namely—cold chain increases the profits of the store has been rated high on a scale of 5 by the respondents. It has got a mean value of 4.31. The mean value of the other dependent variable 'Cold chain increases the footfall in retail store' is also promising and very close to the first dependent variable (4.29).

Impact of various benefits of a cold chain on overall profitability of a store and customer footfall in the store It has been found and discussed that which

Table 1.13 Consolidate means and standard deviation for statements related to the benefits of cold chain

Sl. no.	Benefits of cold chain management to the store	Mean	Std. deviation
1	It maintains the quality of the products	4.19	0.94
2	It is very important for storage of vegetables	3.88	0.85
3	It helps in restoring the texture of the products	3.27	1.2
4	It helps in protecting the products from bacteria	**4.45**	0.57
5	It helps in retaining the nutrients of the products	**4.41**	0.53
6	It helps in retaining the colour of the products	3.88	0.72
7	It helps in retaining the odour of the products	4.15	0.73
8	It helps in increasing the storage and shelf life of the product	**4.33**	0.56
9	It helps in protecting the products from humidity	**4.44**	0.64
10	It helps in minimizing the spoilage and wastage	**4.38**	0.59
11	It helps in retaining the freshness of the products	**4.31**	0.59
12	It helps in retaining the moisture of the products	**4.53**	0.55
13 (DV1)*	**An effective cold chain increases the overall profits of a retail store**	4.32	0.57
14 (DV2)*	**An effective cold chain increases the footfall in a retail store**	4.29	0.88

*DV1 and DV2 stand for the dependent variables. The rest 12 variables are independent variables. These will be used further to find out that which variables (independent variables) impact the profitability and footfall

variables under a cold chain affect the profit and 'footfall'. To check the significance of those variables, multiple regression (stepwise) has been applied with dependent variables:

1. **An Effective Cold chain increases the Overall Profits of a retail Store** and
2. **An Effective Cold chain is Mandatory for good footfall in a retail store.**

The list of null and alternative hypotheses regarding impact of benefits of a cold chain on overall profits is presented here. All alternative hypotheses are directional hypotheses. These show a positive direction (positive impact) of various benefits of the cold store on overall profits of a retail store.

3 List of Null and Alternative Hypotheses Regarding Impact of Benefits of a Cold Chain on Overall Profits

Ho 1 Maintaining quality of the products has no impact on the overall profits of a retail store.

Ha 1 Maintaining quality of the products has a positive impact on the overall profits of a retail store.
Ho 2 Storage of vegetables has a no impact on the overall profits of a retail store.
Ha 2 Storage of vegetables has a positive impact on the overall profits of a retail store.
Ho 3 Restoration of the texture of the products has no impact on the overall profits of a retail store.
Ha 3 Restoration of the texture of the products has a positive impact on the overall profits of a retail store.
Ho 4 Protection of the products from bacteria has no impact on the overall profits of a retail store.
Ha 4 Protection of the products from bacteria has a positive impact on the overall profits of a retail store.
Ho 5 Retention of the nutrients of the products has no impact on the overall profits of a retail store.
Ha 5 Retention of the nutrients of the products has a positive impact on the overall profits of a retail store.
Ho 6 Retention of the colour of the products has no impact on the overall profits of a retail store.
Ha 6 Retention of the colour of the products has a positive impact on the overall profits of a retail store.
Ho 7 Retention of the odour of the products has no positive impact on the overall profits of a retail store.
Ha 7 Retention of the odour of the products has a positive impact on the overall profits of a retail store.
Ho 8 Increase in storage and shelf life of the product has no impact on the overall profits of a retail store.
Ha 8 Increase in storage and shelf life of the product has a positive impact on the overall profits of a retail store.
Ho 9 Minimization of the spoilage and wastage has no impact on the overall profits of a retail store.
Ha 9 Minimization of the spoilage and wastage has a positive impact on the overall profits of a retail store.

4 Test for Multicollinearity

Multicollinearity is a phenomenon in which two or more predictor variables (independent variables) in a multiple regression model are highly correlated, meaning that one can be linearly predicted from the others with a substantial degree of accuracy. Multicollinearity is checked with the help of VIF value. VIF stands for variance inflation factor. VIF is calculated with the help of SPSS in this study, and the variables with more than 3 VIF value are dropped. VIF is calculated by keeping one independent variable as dependent variable and rest of the independent variables. This process is repeated till **all the independent variables are considered as dependent variables keeping the rest of the variables independent**.

Each variable played as dependent and independent for calculation of cross VIF. All the values of VIF were below 3, which showed that there was no multicollinearity among the various independent variables.

5 a Dependent Variable: Effective Cold Chain Increases the Profits of Retail Store

The coefficient Table 14 shows which variable out of the total 12 independent variables significantly affects the dependent variable. There are five benefits, namely **freshness of the products, increase in storage and shelf life of the product, protection of the products from bacteria, minimization of the spoilage and wastage and maintenance of the quality of the products** affect the profitability of the stores significantly.

The list of excluded variables or the variables which do not have significant impact of profits of retail stores are listed in Table 15.

Table 14 Coefficients (a)

Model		Unstandardized coefficients		Standardized coefficients	*T*	Sig.
		B	Std. error	Beta		
5	(Constant)	−0.568	0.297		−1.908	0.059
	It helps in retaining the freshness of the products	0.303	0.065	0.309	4.675	0.000
	It helps in increasing the storage and shelf life of the product	0.241	0.068	0.236	3.537	0.001
	It helps in protecting the products from bacteria	0.234	0.064	0.232	3.665	0.000
	It helps in minimizing the spoilage and wastage	0.241	0.069	0.247	3.488	0.001
	It maintains the quality of the products	0.106	0.031	0.175	3.413	0.001

Table 15 Excluded variables (f)

Model		Beta in	*T*	Sig.
5	It is very important for storage of vegetables	0.087 (e)	1.625	0.107
	It helps in restoring the texture of the products	0.029 (e)	0.510	0.611
	It helps in retaining the nutrients of the products	0.152 (e)	1.908	0.059
	It helps in retaining the colour of the products	0.069 (e)	1.217	0.226
	It helps in retaining the odour of the products	0.023 (e)	0.351	0.726
	It helps in protecting the products from humidity	0.089 (e)	1.237	0.219
	It helps in retaining the moisture of the products	0.115 (e)	1.466	0.146

e Predictors in the model: (constant), **freshness of the products, increase in storage and shelf life of the product, protection of the products from bacteria, minimization of the spoilage and wastage and maintenance of the quality of the products**
f Dependent variable: **effective cold chain increases the profits of retail store**

6 Results of the Hypotheses Testing

Ho 1 Maintaining quality of the products has no impact on the overall profits of a retail store.
Ha 1 Maintaining quality of the products has a positive impact on the overall profits of a retail store.

The value under significant column of Table 14 is 0.01 which is less than the value of '*p*' at 5% significance level (0.05). Hence, null hypothesis is rejected and alternate hypothesis is accepted. Hence, it may be concluded that maintaining quality of the products has a positive impact on the overall profits of a retail store.

Ho 2 Storage of vegetables has a no impact on the overall profits of a retail store.
Ha 2 Storage of vegetables has a positive impact on the overall profits of a retail store.

The value under significant column of Table 15 is 0.107 which is more than the value of '*p*' at 5% significance level (0.05) that is why, the null hypothesis is accepted. Hence, it may be concluded that simply storage of vegetables has no impact on the overall profits of a retail store.

Ho 3 Restoration of the texture of the products has no impact on the overall profits of a retail store.
Ha 3 Restoration of the texture of the products has a positive impact on the overall profits of a retail store.

The value under significant column of Table 15 is 0.611 which is more than the value of '*p*' at 5% significance level (0.05) that is why, the null hypothesis is accepted. Hence, it may be concluded that restoration of texture has no impact on the overall profits of a retail store.

Ho 4 Protection of the products from bacteria has no impact on the overall profits of a retail store.

Ha 4 Protection of the products from bacteria has a positive impact on the overall profits of a retail store.

The value under significant column of Table 14 is 0.00 which is less than the value of '*p*' at 5% significance level (0.05). Hence, null hypothesis is rejected and alternate hypothesis is accepted. Hence, it may be concluded that protecting products from bacteria has a positive impact on the overall profits of a retail store.

Ho 5 Retention of the nutrients of the products has no impact on the overall profits of a retail store.

Ha 5 Retention of the nutrients of the products has a positive impact on the overall profits of a retail store.

The value under significant column of Table 15 is 0.059 which is more than the value of '*p*' at 5% significance level (0.05) that is why, the null hypothesis is accepted. Hence, it may be concluded that retention of the nutrients of products has no impact on the overall profits of a retail store.

Ho 6 Retention of the colour of the products has no impact on the overall profits of a retail store.

Ha 6 Retention of the colour of the products has a positive impact on the overall profits of a retail store.

The value under significant column of Table 15 is 0.226 which is more than the value of '*p*' at 5% significance level (0.05) that is why, the null hypothesis is accepted. Hence, it may be concluded that retention of the colour of the product has no impact on the overall profits of a retail store.

Ho 7 Retention of the odour of the products has no positive impact on the overall profits of a retail store.

Ha 7 Retention of the odour of the products has a positive impact on the overall profits of a retail store.

The value under significant column of Table 15 is 0.726 which is more than the value of '*p*' at 5% significance level (0.05) that is why, the null hypothesis is accepted. Hence, it may be concluded that retention of the odour of the products has no impact on the overall profits of a retail store.

Ho 8 Increase in storage and shelf life of the products has no impact on the overall profits of a retail store.

Ha 8 Increase in storage and shelf life of the products has a positive impact on the overall profits of a retail store.

The value under significant column of Table 14 is 0.01 which is less than the value of '*p*' at 5% significance level (0.05). Hence, null hypothesis is rejected and alternate hypothesis is accepted. Hence, it may be concluded that increase in storage and shelf life of the product has a positive impact on the overall profits of a retail store.

Ho 9 Minimization of the spoilage and wastage has no impact on the overall profits of a retail store.

Ha 9 Minimization of the spoilage and wastage has a positive impact on the overall profits of a retail store.

The value under significant column of Table 14 is 0.01 which is less than the value of '*p*' at 5% significance level (0.05). Hence, null hypothesis is rejected and alternate hypothesis is accepted. Hence, it may be concluded that minimization of the spoilage and wastage has a positive impact on the overall profits of a retail store.

7 Benefits of Cold Chain in Retail Stores and Its Impact on Footfall in the Store

The impact of independent variables (benefits of the effective management of cold chain) has also been checked on the footfall of the customers in a retail store. In this case also, stepwise multiple regression was applied. The results have discussed ahead. The following hypotheses have been formed:

8 List of Null and Alternate Hypotheses Created for Impact on Footfall in the Store

8.1 Hypotheses

Ho 1 Maintaining quality of the products has no impact on the footfall of the retail store.

Ha 1 Maintaining quality of the products has a positive impact on the footfall of the retail store.

Ho 2 Storage of vegetables has a no impact on the footfall of the retail store.

Ha 2 Storage of vegetables has a positive impact on the footfall of the retail store.

Ho 3 Restoration of the texture of the products has no impact on the footfall of the retail store.

Ha 3 Restoration of the texture of the products has a positive impact on the footfall of the retail store.

Ho 4 Protection of the products from bacteria has no impact on the footfall of the retail store.

Ha 4 Protection of the products from bacteria has a positive impact on the footfall of the retail store.

Ho 5 Retention of the nutrients of the products has no impact on the footfall of the retail store.

Table 16 Model summary

Model	*R*	*R* square	Adjusted *R* square	Std. error of the estimate
1	0.666 (a)	0.444	0.439	0.46305
2	0.737 (b)	0.543	0.534	0.42194
3	0.754 (c)	0.568	0.556	0.41200
4	**0.766 (d)**	**0.587**	**0.571**	**0.40492**

d Predictors: (constant)

Ha 5 Retention of the nutrients of the products has a positive impact on the footfall of the retail store.

Ho 6 Retention of the colour of the products has no impact on the footfall of the retail store.

Ha 6 Retention of the colour of the products has a positive impact on the footfall of the retail store.

Ho 7 Retention of the odour of the products has no positive impact on the footfall of the retail store.

Ha 7 Retention of the odour of the products has a positive impact on the footfall of the retail store.

Ho 8 Increase in storage and shelf life of the product has no impact on the footfall of the retail store.

Ha 8 Increase in storage and shelf life of the product has a positive impact on the footfall of the retail store.

Ho 9 Minimization of the spoilage and wastage has no impact on the footfall of the retail store.

Ha 9 Minimization of the spoilage and wastage has a positive impact on the footfall of the retail store.

The model summary, as presented in Table 16, shows that four steps were taken for the optimization of the model and results. The fourth step shows value of *R* square as 0.587 which finds that around 60% of the variance is explained by the model, which is quite significant. The value of adjusted *R* square is bit less; however, that too is sufficient.

Protection of product from bacteria, retaining the freshness of the products, maintains the quality of the products and protecting the products from humidity.

9 e Dependent Variable: Effective Cold Chain Increases the Footfall of Customer in the Retail Store

Table 17 shows the ANOVA statistics. Under the model column—5 represents the values which have come out at the fourth step. The ANOVA Table 17 tells whether the dependent variable is significantly affected by the independent variable. In the Sig.

Table 17 ANOVA (e)

Model		Sum of squares	Df	Mean square	*F*	Sig.
4	Regression	24.885	4	6.221	37.944	0.000 (d)
	Residual	17.544	107	0.164		
	Total	42.429	111			

d Predictors: (constant)

column, the value is 0.000, which is less than the critical value of *p*—0.05. Hence, there is a significant impact of independent variables on dependent variable. With the value in Sig. column (d) is also added with 0.000, which means (see Table 17 footnote) that it is significant when the dependent variable is predicted by the four independent variables namely—protection of product from bacteria, retaining the freshness of the products, maintaining the quality of the products and protecting the products from humidity.

The coefficient Table 18 shows which variable out of the total 12 independent variables significantly affects the dependent variable.

There are four benefits, namely protection of product from bacteria, retaining the freshness of the products, maintaining the quality of the products and protecting the products from humidity which affect the footfall of customers in the retail store. The list of excluded variables or the variables which do not have significant impact of profits of retail stores are listed in Table 19.

Table 18 Coefficients (a)

Model		Unstandardized coefficients		Standardized coefficients	*T*	Sig.
4	(Constant)	−0.191	0.373		−0.514	0.608
	It helps in protecting the products from bacteria	0.459	0.084	0.421	5.499	0.000
	It helps in retaining the freshness of the products	0.280	0.084	0.265	3.338	0.001
	It maintains the quality of the products	0.114	0.041	0.174	2.759	0.007
	It helps in protecting the products from humidity	0.177	0.081	0.184	2.194	0.030

a Dependent variable: **effective cold chain increases the footfall of customer in the retail store**

Table 19 Excluded variables (e)

Model		Beta in	*T*	Sig.
4	It is very important for storage of vegetables	0.083 (d)	1.246	0.215
	It helps in restoring the texture of the products	−0.052 (d)	−0.761	0.448
	It helps in retaining the nutrients of the products	0.104 (d)	1.112	0.269
	It helps in retaining the colour of the products	0.007 (d)	0.102	0.919
	It helps in retaining the odour of the products	0.082 (d)	1.111	0.269
	It helps in increasing the storage and shelf life of the product	0.127 (d)	1.659	0.100
	It helps in minimizing the spoilage and wastage	0.117 (d)	1.393	0.167
	It helps in retaining the moisture of the products	0.146 (d)	1.460	0.147

d Predictors in the model: (constant), protection of product from bacteria, retaining the freshness of the products, maintains the quality of the products and protecting the products from humidity
e Dependent variable: **effective cold chain increases the footfall of customer in the retail store**

10 Results of the Hypotheses Testing

Ho 1 Maintaining quality of the products has no impact on the footfall of the retail store.
Ha 1 Maintaining quality of the products has a positive impact on the footfall of the retail store.

The value under significant column of Table 18 is 0.007 which is less than the value of ‘*p*’ at 5% significance level (0.05). Hence, null hypothesis is rejected and alternate hypothesis is accepted. Hence, it may be concluded that maintaining of quality of the product has a positive impact on the footfall of a retail store.

Ho 2 Storage of vegetables has a no impact on the footfall of the retail store.
Ha 2 Storage of vegetables has a positive impact on the footfall of the retail store.

The value under significant column of Table 19 is 0.215, which is more than the value of ‘*p*’ at 5% significance level (0.05) that is why, the null hypothesis is accepted. Hence, it may be concluded that restoration of the simply storage of vegetables has no impact on the footfall of a retail store.

Ho 3 Restoration of the texture of the products has no impact on the footfall of the retail store.
Ha 3 Restoration of the texture of the products has a positive impact on the footfall of the retail store.

The value under significant column of Table 19 is 0.448 which is more than the value of ‘*p*’ at 5% significance level (0.05) that is why, the null hypothesis is accepted. Hence, it may be concluded that restoration of the restoration of texture has no impact on the footfall of a retail store.

Ho 4 Protection of the products from bacteria has no impact on the footfall of the retail store.

Ha 4 Protection of the products from bacteria has a positive impact on the footfall of the retail store.

The value under significant column of Table 18 is 0.000 which is less than the value of '*p*' at 5% significance level (0.05). Hence, null hypothesis is rejected and alternate hypothesis is accepted. Hence, it may be concluded that protection of products from bacteria has a positive impact on the footfall of a retail store.

Ho 5 Retention of the nutrients of the products has no impact on the footfall of the retail store.

Ha 5 Retention of the nutrients of the products has a positive impact on the footfall of the retail store.

The value under significant column of Table 19 is 0.269 which is more than the value of '*p*' at 5% significance level (0.05) that is why, the null hypothesis is accepted. Hence, it may be concluded that retention of nutrients of the products has no impact on the footfall of a retail store.

Ho 6 Retention of the colour of the products has no impact on the footfall of the retail store.

Ha 6 Retention of the colour of the products has a positive impact on the footfall of the retail store.

The value under significant column of Table 19 is 0.919 which is more than the value of '*p*' at 5% significance level (0.05) that is why, the null hypothesis is accepted. Hence, it may be concluded that retention of the colour of the product has no impact on the footfalls of a retail store.

Ho 7 Retention of the odour of the products has no positive impact on the footfall of the retail store.

Ha 7 Retention of the odour of the products has a positive impact on the footfall of the retail store.

The value under significant column of Table 19 is 0.269 which is more than the value of '*p*' at 5% significance level (0.05) that is why, the null hypothesis is accepted. Hence, it may be concluded that retention of the odour of the products has no impact on the footfall of a retail store.

Ho 8 Increase in storage and shelf life of the product has no impact on the footfall of the retail store.

Ha 8 Increase in storage and shelf life of the product has a positive impact on the footfall of the retail store.

The value under significant column of Table 19 is 0.1 which is more than the value of '*p*' at 5% significance level (0.05) that is why, the null hypothesis is accepted. Hence, it may be concluded that increase in storage and shelf life of the product has no impact on the footfall of a retail store.

The value under significant column of Table 18 is 0.03 which is less than the value of '*p*' at 5% significance level (0.05). Hence, null hypothesis is rejected and alternate hypothesis is accepted. It may be concluded that protection of the products from humidity has a positive impact on the footfall of a retail store.

Ho 9 Minimization of the spoilage and wastage has no impact on the footfall of the retail store.

Ha 9 Minimization of the spoilage and wastage has a positive impact on the footfall of the retail store.

The value under significant column of Table 19 is 0.167 which is more than the value of '*p*' at 5% significance level (0.05) that is why, the null hypothesis is accepted. Hence, it may be concluded that minimization of the spoilage and wastage of products has no impact on the footfall of a retail store.

11 Conclusion

The present study is on effectiveness of the cold chain in retail stores in the Rajasthan state. Rajasthan has varied climatic conditions and produces a variety of fruits and vegetables. Hence, the stores may buy a wide variety of goods locally, store them and sell them by using their cold chains as and when required.

This was found that the maximum space of the cold chains is filled with the milk and dairy products and beverages. Lesser space is given to the ready-to-eat and ready-to-cook products. There are mainly five benefits that affect the profitability of a retail store significantly. These factors are—**freshness of the products, increase in storage and shelf life of the product, protection of the products from bacteria, minimization of the spoilage and wastage and maintenance of the quality of the products**.

An effective cold chain also affects the footfall. For instance, everyone wants a fresh product—fruit or vegetable, properly frozen ready-to-eat products and chilled beverage. That is why, customers are more attracted towards the store which maintains an effective cold chain. It may be concluded from the study that **protection of product from bacteria, retaining the freshness of the products, maintaining the quality of the products and protecting the products from humidity which affect** the footfall of customers in the retail store are those five benefits of a cold chain which increases the footfall in a cold chain. Technically speaking, the above benefits have significant impact on customer footfall in a retail store.

12 Suggestions and Recommendations

It has been found in the study that major benefits of cold chain includes the better shelf life and protection of products so that the quality of the products can be maintained.

Further, an open-ended question was also asked from the respondents to give their suggestions regarding the effective cold chain management. Consolidating all, the following recommendations have been made:

1. Technically trained staff is the need of hour in cold chains. The cold chain management has become very technical job due to the wide variety of products, limited space and rising electricity and carrying cost of stock. Hence, the staff who is managing the cold chain must be effective so that they can use their skills to optimize the operations in a cost-effective manner.
2. There should be an electricity backup so that the wastage and spoilage can be minimized. Further, effective electricity management is also needed to reduce the costs at the same time.
3. Review, audit and maintenance of the whole system must be a continuous process so that sudden breakdowns in technical infrastructure may be avoided.
4. Prompt and full communication regarding the supply chain should be made available to the store managers so that they may arrange the cold chains accordingly by making space for new products and/or retaining the products for peak time of demand.
5. Cold chain managers should maintain the in-flow and out-flow data on daily basis from their cold chains so that a computerized database may be maintained which may further help them in predicting their future demand.
6. This system of RFID in the cold chain can bring a boom in the processed food industry. The places near Jaipur and its periphery are rich in milk products, vegetables, fruits, etc. These are being processed and are frozen in ready-to-use form being marketed for the customers. The business must adopt early the functions of RFID in the industry which is still at an early stage. A cold chain which not only refers to the transportation of temperature sensitive products along a supply chain but also to be managed through thermal and refrigerated packaging methods in the logistical planning for the protection of the integrity of these shipments.

References

Bharti, A. (2014). Examining market challenges pertaining to cold chain in the frozen food industry in Indian retail sector Apeejay. *Journal of Management Sciences and Technology, 2*(1), 33. ISSN 2347-5005.

Bharti, A. (2016). Supply chain management and strategy implementation for perishable goods. In N. Kamath & S. Saurav (Eds.), *Handbook of research on strategic supply chain management in the retail industry* (pp. 152–169). Hershey, USA: IGI Global.

Bharti, A. (2017). *Effective management of cold chains for frozen food products: A study of selected retail outlets* (PhD thesis). NIMS University, Jaipur, Rajasthan.

Bharti, A., & Goyal, A. (2017). Managing various aspects of cold chain management for overall profitability in the selected retail stores of Rajasthan. *BRDU International Journal of Multidisciplinary Research, 2*(2), 119–130. ISSN 2455-278X.

Beardsell, D., Francis, J., & Ridley, D. (2002). Health-promoting constituents in plant-derived edible oils. *Journal of Food Lipids, 9*(1), 1–34.

Boronico et al. (1996). Customer service: The distribution of seasonal food products under risk. *International Journal of Physical Distribution & Logistics Management, 26*(1), 25–39.

Henderson. (1994). Logistics: The key to success for the Australian food industry push into Asia. *The International Journal of Logistics Management, 5*(2), 63–70.

Joshi, R., Banwet, D. K., & Shankar, R. (2009). Indian cold chain, modeling the inhibitors. *British Food Journal, 111*(11), 1260–1283.

Kelepouris, T., Pramatari, K., & Doukidis, G. (2007). RFID-enabled traceability in the food supply chain. *Industrial Management & Data Systems, 107*(2), 183.

Kovacs, G. (2004). Corporate environmental responsibility in the supply chain. *Journal of Cleaner Production, 16*, 1571–1578.

Lisa, K. (2013). *Use of cold chains for reducing food losses in developing countries* (PEF White Paper No. 13-03). The Postharvest Education Foundation (PEF).

Liu et al. (1999). Co-ordination of international channel relationships: Four case studies in the food industry in China. *Journal of Business & Industrial Marketing, 14*(2), 130–150.

Manning, L., Baines, R. N., & Chadd, S. A. (2006). Food safety management in broiler meat production. *British Food Journal, 108*(8), 605–621.

Manos, B., & Manikas, I. (2010). Traceability in the Greek fresh produce sector: Drivers and constraints. *British Food Journal, 112*(6), 640–652.

Maxwell et al. (2006). Functional and systems aspects of the sustainable product and service development approach for industry. *Journal of Cleaner Production, 14*, 399–416.

Nielsen, A. C. (2006). *Economic climate boosts FMCG spend.* Available at http://www.bizcommunity.com/PressOffice/PressRelease.aspx?i=358&ai=10907.

Salin et al. (2003). A cold chain network for food exports to developing Bo Wei, June 20th, 2011 38 countries. *International Journal of Physical Distribution & Logistics Management, 33*(10), 918–933.

Viswanadham, N. (2005). *Cold chain management: India-Singapore initiative*. Singapore: The Institute for South Asian Studies, National University of Singapore.

Websites

http://ijmdr.in/data/documents/FEBRUARY__PAPER-14-1.PDF.

http://trade.gov/topmarkets/pdf/Cold_Chain_india.pdf, 2012.

http://www.ediindia.org/doc/SpecialPDF/chp-1.pdf, 2014.

http://www.ediindia.org/doc/SpecialPDF/chp-1.pdf, 2010.

http://www.fnbnews.com/Top-News/cold-chain-industry-in-india–present-status-and-future-prospects-38621.

http://www.fnbnews.com/Top-News/Indian-processed-food-industry-5th-in-terms-of-export-expected-growth, Indian processed food industry 5th in terms of export, expected growth, Vinu Shankar, Thursday, 29th January, 2015, 08:00 (IST).

http://www.nccd.gov.in/PDF/Mofpi.pdf.

https://seafood.oregonstate.edu/sites/agscid7/files/snic/managing-the-cold-chain-for-quality-and-safety.pdf.

http://www.thehindu.com/todays-paper/tp-national/tp-newdelhi/rajasthan-has-huge-potential-to-become-food-processing-hub/article4423687.ece, 2015.

Block Chain

Block Chain and Its Application

Dheeraj Kumar, Ravi Kant Singh, and Apurba Layek

Abstract In this research article, the investigation has been made possible to gather some ideas about Block Chain technology and its recent applications. It has also been tried to explore some other implementations of the related protocols. At the initial stage, discussions on the Bitcoin (the first-ever application of the technology) are being provided. Block Chain technology is known as the technological basis on which Bitcoin is built. This technology has made it possible that every kind of transaction is executed in a proper manner of decentralized way. It does not require any trusted kind of the third party in between the whole transaction. By offering a distributed, undisputable, public verified records of the transactions, the Block Chain technology potentials to renovate many industries. In this chapter, a critical and detailed survey has been performed aiming toward the area of application of the Block Chain technology, its impact on the society, and its possible use cases. The idea purpose of this chapter is to provide that much ability to the readers to familiarize and to understand the state of the art of the Block Chain world, in both the society and technology also.

Keywords Block Chains · Bitcoins · Smart contracts · Business application · Cryptocurrencies

1 Introduction

We all human beings have a normal tendency toward the technologies to currently granted were quiet revolutions in their time. Admit ourselves that up to what proportion the smartphones have modified the manner to live and work. It accustomed be that once individuals were out of the workplace, they were gone, as a result of a phone was tied to an area, to not someone. Currently, we have got world nomads building new businesses traditional from their phones. Furthermore, to think: as a result, smartphones are around us for merely a decade (Kesharwani 2018).

D. Kumar (✉) · R. K. Singh · A. Layek
Department of Mechanical Engineering, National Institute of Technology Durgapur, Durgapur, West Bengal 713209, India
e-mail: dknitdgp1@gmail.com

K. Kumar and J. P Davim (eds.), *Supply Chain Intelligence*, Management and Industrial Engineering, https://doi.org/10.1007/978-3-030-46425-7_6

We are in the midst of the global age revolution. *Block Chain*, a distributed information database that keeps and maintains the overall growing datasheet records in a proper manner called "blocks." Let us discuss the past ten years, actually what happened (Butum et al. 2018).

The existence of the Block Chain is being visualized in the economic crisis of 2008. It has the foundations of itself in the research field into cryptocurrencies. In other words, it can also illustrate as currencies are denominated, and entire digital technology is managed. In that respect, this technology brings a revolutionary environment that enables the development rate fast and implementation of the Internet of the transaction (IoT) (Zheng et al. 2017). In a system, we find ourselves trusted, but without trusted third parties. How paradoxical! And how evolutionary!

2 Toward a Definition

Block Chains itself contain many definitions, according to the specific application types and extremely technical. Coinbase organization, largest exchange company of cryptocurrency, defines it as "*a distributed information, publicly available ledger which contains the overall transaction history of every Bitcoins*." Some broaden been brought by the Oxford English Dictionary for Block Chains as "*a fully digitalized ledger in which the all transaction is made in the Bitcoin.*" These two definitions highlight the benefits of the Block Chains as served as a digital ledger. Most of the literature is supporting this definition. In theory, Block Chain is a duplicate register which is sharing the information between the nodes of a network (Shaikh and Lashari 2017).

Somewhat a few broader definitions are being presented in the secure. Here Block Chains are defined as "*some special type of structured data which enables us to identify and to track the transactions in a digitalized manner.*" It also enables the system to be transparent by sharing the entire information database via a distributed network of computers. It provides a trusted network of distributed forms. This platform of distributed ledger offers a transparent and secure means of tracking the ownership and assets transfer (Sarmah 2018).

Block Chains itself need some more concise and precise definition. Based on the theoretical aspects of Block Chain technology, it has been defined the Block Chain.

> a decentralized database system itself containing sequential information, cryptographically linkages of the digitalized signature assets transaction, all of this governed by consensus model.

This concise definition of Block Chains will permit the application from the broad swath of industries and use cases for the Block Chain in different industries.

3 Block Chains: A Brief History

- *Bitcoin* was the first significant innovations, sometimes known as digital currency experiments. In the current situation, market cap of Bitcoin lies in between $10 and $20, which is owned by many of the peoples.
- *Block Chain* was the second most innovative findings in this field. It brought the realization about the underlying technology, which was currently using the Bitcoin. It senses that Bitcoins can be separated from the currency and can be used for all kinds of inter-organizational cooperation. Now ongoing research topic is running in most of the financial institutions about Block Chains (Zheng et al. 2017).
- A *smart contract* was the third one innovation in the financial market. It builds computer programming directly into the second generation Block Chain system. So that permission can be granted for the financial instruments to be represented.
- *Proof of stake* is the fourth significant innovations, sometimes often called Block Chain thinking. In the present scenario, the Block Chains are secured by "proof of work." In this group, the decision making is done by the total computer power. This group of computing is called "miners."
- *Block Chain scaling* is the fifth primary concept on the horizon. Scaled Block Chain enhances the speed of the network processes for every transaction. It is expected to be fast enough to accelerate the IoT and go head to head with the payments.

4 Why so Much Excitement Around Block Chain Concept?

Let us think of an introductory situation if anyone can transact indirect way peer to peer having full confidence, without any interference of intermediaries, and without the occurrence of third-party trusty. Then, it becomes to make secure transactions away from skim wealth off (Uber, Airbnb). It can be easily seen a huge number of examples that day per day investment is increasing at a faster rate, increasing articles, conferences.

Another main reason for this kind of eagerness is the scalability of this growing technology. It is that much true that "proof to work" is the basic principle opted by the Block Chain technology. It can be vulnerable to computer viruses attacking the "Goldfinger" type. The characteristics of this attack are to gather the computational power to some specific network, which is related to a hundred attacks daily. Once this information is being achieved, the attacker gets that much free space to validate or invalidate certain transactions (Ben Ayed and Belhajji 2018). The purpose of such kind of attacks is to the nervousness of the user's confidence in this Block Chain technology ('Www.Cybrosys.Com Www.BlockChainexpert.Uk1').

5 Fundamental Principles of the Block Chains

The Block Chain creates a digitalized trusted platform, with the confirmation that once the information has been inputted, it is almost impossible to tamper it. It has the degree of freedom to verify and trace every step for all the actors involved in it. There is always a new transaction for each time, creation of the new blocks for each step, and attached to the existing one blocks; thus, it is called as the "Block Chain." At the same time, all the blocks are updated on a shared network. It has the advantage that it gives complete information about the transactions involved during the entire process. Thus, it maintains a chain of integrity, and sometimes it is often called a chain of trust (Sultan et al. 2018).

There is another critical aspect of Block Chain technology, termed as *smart contracts*. It includes all the algorithm-based programming in which information is being given to the Block Chain to run automatically in the agreed procedure. The examples of such procedures are transactions and reporting. For the illustration, in a ship, if a container has been loaded, all of its events are encoded in the *bill of lading Block Chain*. After the mechanism, a smart contract established, which helps to pay automatically to the terminal operator. Moreover, it can be concluded that on coming into the picture, smart contracts elaborate the application of Block Chains by providing a unique and enforceable documentation record (Hacker et al. 2019) (Fig. 2).

An overview of Block Chain technology: How it works
See Fig. 1.

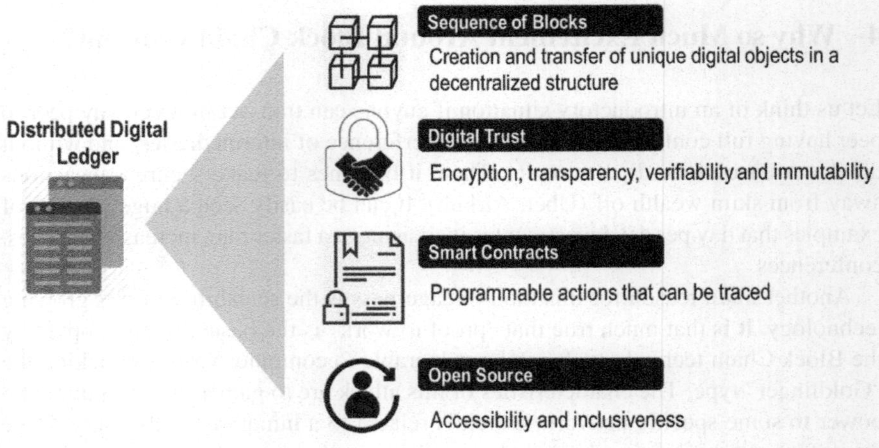

Fig. 1 How Block Chain works

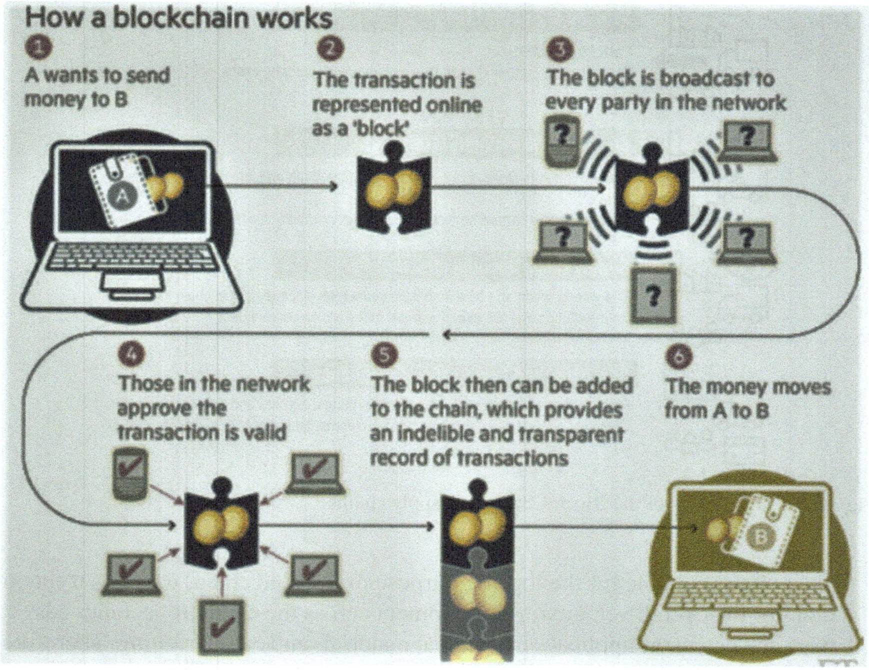

Fig. 2 Core principles of the Block Chains

6 Benefits of the Block Chain Technology

The application of the Block Chain to the supply chain includes itself with a series of expected ***benefits***:

1. **Speed of the supply chains** This is notably the result of quicker transactions, like payments, that area unit a standard reason behind delays. There is less latency within the system, benefiting the income and reducing inventory carrying prices.
2. **Visibility of the supply chains (tracking)** Because of its structure distributed in nature, a Block Chain is extraordinarily troublesome to tamper with, and every dealing transaction desires the validation of the concerned actors. It is so potential to use the Block Chain to trace shipments on an intermodal transport chain, determine problems inflicting delays, and coordinate activities. A crucial facet issues a symptom of delivery that permits the ultimate settlement of the dealing. Due to its potential visibility, a Block Chain is familiar to produce a market wherever transport and supply service suppliers may bid to hold or handle the consignment related to specific blocks (Minis et al. 2011).
3. **Security of the supply chains (tracing)** Since it is the real fact that every event of the supply chain is being kept within the Block Chain, as they acquire someplace. It becomes easy to look where, how, and when the specific work or event is taking

Fig. 3 Feasible profits of the Block Chains on supply chains

place. It fits suitable for the logistic purpose of the cold chain, which is trying to monitor the overall veracity of the shipment across the chain. It becomes easy to trace and detect the application of subpar materials and counterfeiting. Therefore, information of the product regarding the origination and batch number can be included (Chandra and Grabis 2016).

4. **Standards and agreement of certificates** for the agreement of the standards and guidelines of the certificates, Block Chain provides an easy and productive path. Events of the Block Chains prove that the cargo has been handled with proper specific modes, distribution, and carrier centers. The sequence followed during the event can give the right information about the energy uses, impact on the environment (Emissions of the carbon dioxide) (Fig. 3).

It can be concluded that after improving all the aspects, the integration level of the supply chain rises substantially.

Last but not least, one important benefit of the Block Chain application is related to the ***"hacking and integrity of the network."***

Although it is almost impossible to hack specific applications theoretically in the upper level, which is connected to Block Chain infrastructure, they have recently allowed the hackers to steal the substantial quantity of cryptocurrencies, arising the question on the integrity of the Block Chains (Bruyn 2017).

7 Core Categories of Block Chain Usage

The usage of the Block Chain a massive into four major sections. These sections are shortly related to keeping records on the one side and another side facilitating the transactions.

Static registry All the information and references are stored into a distributed info (ledger); it has to be immutable and verifiable. For transportation purposes, this may involve all the record of quality possession (automobiles, transports) and registration data. Static Block Chains tend to be less process-intensive since records are not frequently modified.

Smart contracts A distributed database (info) that contains recorded circumstances desired to trigger once when the action is just about to complete. The actions are such as payment of bills or the transfer of an asset. Operations related to transportations are prone to such smart contracts as fare collections, insurance claims, and bill payments. For example, if a delivery enters a terminal or if a parcel is being delivered, then automatic payments can be made possible, and the transactions are considered as complete dealings (Sarmah 2018).

Dynamic registry It is quite similar to a static registry; however, during this case, the database info is updated repeatedly as supplementary data nowadays and as resources are changed. A supply chain may be a relevant example of a dynamic record that is perpetually updated because of its intensiveness of the associated transactions (Hübnet 2007).

Payment infrastructure A distributed info that supports cryptocurrencies and, therefore, the connected transactions. Merged with smart contracts, cryptocurrencies have the potential to be wont to settle contracts once outlined circumstances (e.g., delivery) on (Madavi 2008) (Fig. 4).

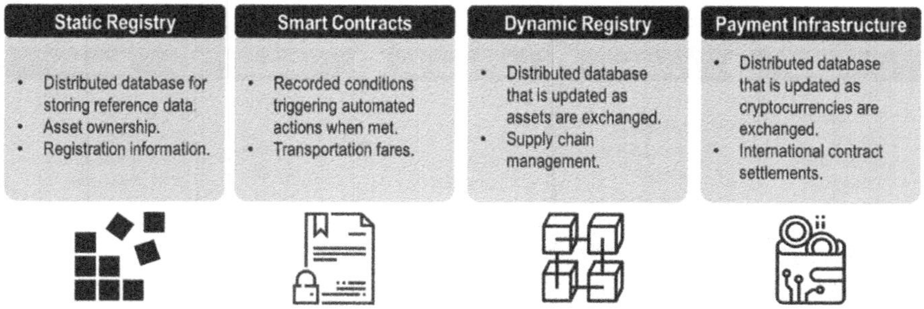

Fig. 4 Core categories of Block Chain usage

8 Value Creation Through Block Chains

The Block Chain provides value to the domain of its logistic chain application. It is known the fact that the proposition of the principal value involves the management of the contract, coordination of the stakeholders (to share information more accurately), and dis-intermediation (so that interaction can be made directly without the presence of the third party). Those value creations are mentioned below:

- **Principles**: This principle involves all the information blocks being stored and encoded on the scattered network. For enhancing the nature of flexibility of the Block Chains, smart contracts are used. Because it enables the capability of automatically sensing and verifying the executed terms of a predefined project (Essaf et al. 2019).
- **Functions**: A Block Chain fulfills various functions relating to the stored information in itself with the distributed info (database). The information implanted in the system cannot be changed without the permission of the stakeholders. The overall information and transactions are to maintained transparency. It means that ledger can be visible for all the members working in the system with allotted permission.
- **Usage of the information**: As the principles and functions are identified earlier, so this section helps the Block Chains to share the information of transactions. It improves the visibility property of those assets, which is going for the transaction since both of the buyer and provider can see the attributes related to those assets such as price, availability, and quantity. An essential feature of this segment enables to evaluate the transaction involved through an audit (Casino et al. 2019).
- **Processes**: Block Chain supports and offers essential features, i.e., contract management. As it is a well-known fact that contracts are visible, shareable, and traceable also. It further gives better coordination of the demand and supply.
- **Outcomes**: Final steps are termed as outcomes, and it is expected that efficiency, effectiveness, and error minimization (resilience) of the process exhibited and supported by the Block Chains will improve (Fig. 5).

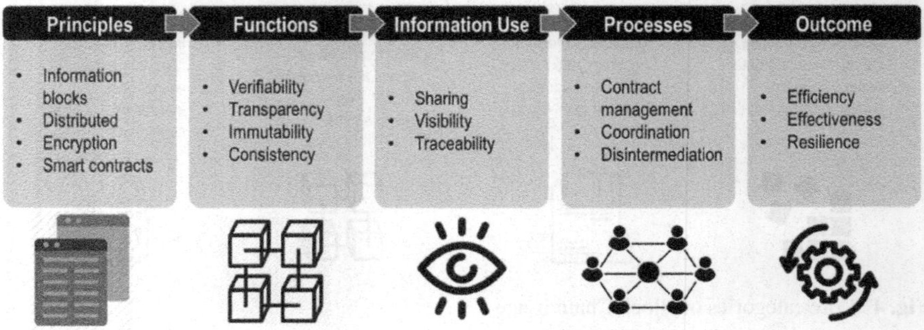

Fig. 5 Block Chains and its value creation

9 Applications of the Block Chains

There are various applications of the Block Chains other than digitalized currency. It has been observed that on coming into existence of smart contracts in Ethereum, it gives a pathway for the application of the Block Chains financially. The most important use cases of the Block Chains are discussed in this section (Rantala 2019).

1. **Financial contracts**

Derivatives of financial contracts are most suited for the Block Chain implementation. The reason behind this fact is that the smart contracts built on underlying assets. This behavior of the assets offers an advantage to the system of triggering the event, which is responsible for the execution of the contracts. It is easily programmed according to the need of the circumstances. It becomes easy to enhance the efficiency and visibility of the global market by adopting the automating the derivatives of the financials. As a result, it also reduces transaction costs. In these days, many of the financial derivatives traded over the counter (OTC); it means that their pricing policy is not transparent. It may lead to allow the local markets making the organization to extract huge demand of fees for their as intermediate financials. The presence of Block Chains cuts the intermediate connections of the financials (Foroglou and Tsilidou 2015).

2. **Asset tracking**

Another use of the Block Chains is tracking the assets for the proof of the right ownership or source of a particular asset. Sometimes, addressing of the goods becomes most important to address the stolen goods or the so-called blood diamonds. It becomes quite necessary to keep the records having the quality of assurance like; it should be publicly valuable, immutable, verified proofs for validation of the ownership at any time. Block Chains provide all the attributes needed to fulfill the same. It also makes an easy path in the case of tracing all the transactions involving any particular type of items, also if the hands of the global supply change (Chain 2013).

3. **Payment system**

Block Chains can be used for the implementation of the payment system in fiat currency. In cryptocurrencies, it is a natural extension of its working ability to manage payments and transactions too.

4. **Digital identity**

As we have seen that Block Chains can be used to track the ownership of the assets and goods, it can also be used to keep the complete identity of the peoples. Let us discuss a situation if our passport is being stored in a Block Chain. The visa that the authority gives us has all our data regarding entry, a departure from countries as Block Chain transactions. It senses that it has the property of the immutable, verified community, decentralized. All the rules are visible to all and fully automated to reduce the chances of human error that may happen at the time entering the information.

5. **Security**

It has preserves that so many recent problems have occurred in several countries related to electoral fraud. For the solution of its part, a start-up FollowMyVote is banking on the digital formation of the whole electoral system. This start-up has been evolved to apply the concept of Block Chain to ensure that voting has been successfully audited and traceable.

6. **Health and pharmaceuticals**

The healthcare sector has also abounded with the potential application of the Block Chains. In pharmaceuticals, industrial sector legitimacy, authenticity, and traceability of clinical results are supreme. BlockRX ensures the traceability of the supply chain with the application of Block Chain technology. Another application of this technology in the medical sector is to keep records of medicines. By digitalizing the entire system records, it entirely becomes possible to transfer the information of the patients from one healthcare professional to the other one. A perfect example of the existing system that we are looking for faxes, medical forms available on the Internet (Sarmah 2018).

7. **Insurance**

Insurance is another interest in Block Chain usage. In this sector, there are many more possibilities, but we are in a little bit of surprise. For the sake of convenience, there is a better example related to; parametric insurance (Rain vow) enables the surety of the party to be compensated automatically, in the case when a specific event occurs, thanks to the smart contracts. In this dynamic way, we can give our thanks to the interconnectivity of objects (IoT). Due to IoT, it is possible to record automatic insurance contracts (Konstantinidis et al. 2018).

9.1 Application of the Block Chain Technology in the Environment of Greek Economic

1. **Shipping management**

Block Chain technology can perform some selected type of service to control the arrival of the containers in the Greek ports. The container received lastly at the port must have some labeling with a cryptographic hash; it must have matched with the last one. The value of the system automatically keeps records and tracks, and therefore, it prevents mistakes and scams. In the case of labor disputes, it can cost heavy for the poor managing port organization. However, a practical example Oakland port international market, which is free from the labor force. It became only possible due to the adoption of the Block Chain application (Foroglou and Tsilidou 2015).

10 Business Application of the Block Chains

- **The value proposition of Block Chain technology**

Since the basic application of the Block Chain technology is to make surety of the transparency and substantiate correction of the transaction of the data across the ecosystem, the potential use of such practical and renowned technology becomes endless. Some of the virtual currencies like Bitcoin, a multitude of other applications of Block Chains, and its related services have been proposed. In making sense of its related application ecosystem. The potential to verify the value proposition of the Block Chains. These proportions are denoted by six elements, as denoted in short form as ATOMIC *(assets, Trust, Ownership, Money, identity, contracts)*. With the help of these all six elements, there be made written in a computer that can enable new services coming to the global market with less price of the transaction and also having faster execution. In this way, it has been seen that the Block Chain technology can remove the barrier of the intermediaries of the business models (Sultan et al. 2018).

The application of Block Chains has been categorized into the four aspects of the spanning like:

1. A platform for the development
2. Utilized as a smart contract
3. As a marketplace
4. Trusted service provider.

- **Block Chain as a development platform**

In the present scenario, for the development of the application of Block Chains, it requires highly skilled. However, the state of the developer of the Block Chain developer is of immature type. Therefore, Block Chain as a service (BaaS) provides a platform such as Microsoft, IBM, for the development of skillsets rapidly on test Block Chains before working in real situation live ones. In these steps, some more examples are there for the process development of Block Chain technology for the secure sharing of data across the industrial networks. They are Xage (for tamper-proof of the ledger records) and Guardtime (for the verification of the transaction data).

- **Block Chain as a smart contract utility**

The programmatic interface to the Block Chains is provided with the smart contracts. The utility of the smart contracts is defined as "*being that much capable of performing the useful functions like; creation, maintenance or augmentation of the digitalized assets value*." The utility part is being captured and stored in the Block Chain function. For the execution of this numeral code, a predetermined condition has to be defined. Third-party authority coming in between transactions is mitigated to the Block Chain instead, disintermediating the in between transaction. A specific instance can be observed in this functional approach of Visa and DocuSign, who works as a partner for proofing the concept of the Block Chain project. It helps in simplifying the management of transactions among the different parties (Sultan et al. 2018).

- **Block Chain as a marketplace**

For generating and adding value for the robust ecosystem, a marketplace is essential. A marketplace in the global world of crypto-economics provides two significant platforms in the form of Block Chains as (i) infrastructure for the payment purpose and (ii) proof for the ownership, sometimes also called digitalized asset tracking. This ability of the marketplaces has enabled peered open markets with no governing authority. Some of the examples of such kind of organizational firm are Open Bazaar, Soma, which provides disintermediate and accessible from any place. In this type of marketplaces, direct interaction between the buyer and the seller can be made possible through smart contracts.

- **Trusted service application**

Finally, Block Chain technology working as a trusted service application gives a facility of end-to-end function by facilitating the unique purpose application. The more generalized use of Block Chains to enable the transaction through all types of the application consists of programmable assets trusted supplier, ownership, money, identity, and contracts termed as *Block Chain 2.0*. There are two perspective views regarding the trusted service application. On the front end, these service applications built Block Chains with the help of smart contracts. It provides disintermediation, such as secure service to the end users. Furthermore, on the back end, these applications reside on Bitcoins and Ethereum, which is not restricted or stopped (Jani 2019).

Furthermore, currently, many companies are providing APIs (application programming interface) nowadays. It allows the developers to build the applications with the help of mechanism and protocols of the Block Chains.

Now, after having a great discussion on the potential applications of the Block Chains, it has been proposed a matrix formation of size (2 × 2). In Fig. 6, it has been elaborated on the view for the mapping industry sectors versus Block Chain scope (public sector vs. private sector vs. hybrid sector) and its access (as service or application). If the terms used here are elaborated in their sense, it means that application refers to the *program designed suitably to perform a specific task to benefit the* while service is for the *transportation of data/information*.

The following criteria have been assessed for the mapping of the horizontal markets on to the matrix:

1. **Access**—Primary function will be based on to transform data (application purpose) or to present the data (for service purpose)?
2. **Scope**—Block Chain application will be restricted globally (public purpose) or permission for enterprises (private purpose)?

To handle these queries, let us talk about a situation as an example of health care and real estate. The healthcare unit is focused on facilitating a secure transit of the

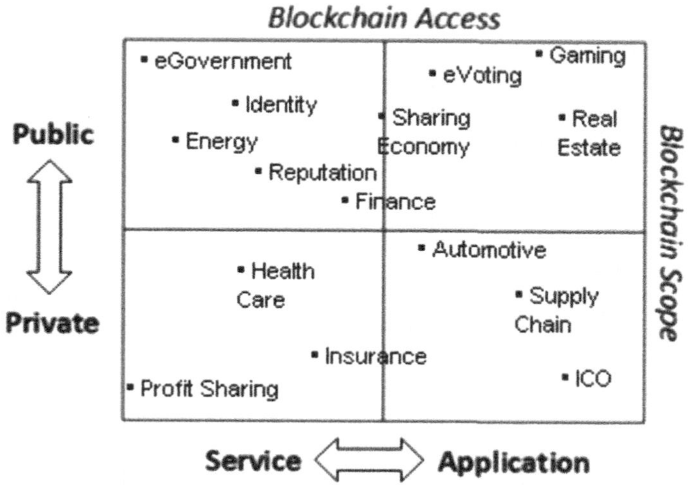

Fig. 6 Scope matrix versus Block Chain access

patient records with the help of Block Chain. So, in this manner, the term *Access* is fore service purposes while scope remains private to partners of the health care. After comparing the real estate view, the industry has shown its interest in knowing the land registration records through Block Chain. So, the result is that this application purpose is only meant to an open and transparent for the public sector (Butum et al. 2018).

Moreover, a global decentralized, trusted value ecosystem can be made possible with the application of Block Chain technology. It can lead to generating new opportunities economically in both the private and public sectors (Kossow and Dykes 2018).

11 Conclusions and Future Work

Block Chain technology is challenging the status of the market transactions (goods, properties, money, asset value) radically, by providing the database in the form of the decentralized platform. In today's business area, Block Chain technology is looking more attractive nature due to its generic nature. However, risks, effects and unintended consequences, Block Chain technology reflected to be an established marketplace. In the business sector, the application of this technology is increasing at a faster rate. A concluding remark comes to the point, after a fruitful discussion that Block Chain has that much potential capability to affect the industries, finance, and cloud computing. Block Chain technology offers a platform for the marketplace to be verified community, immutable distributed records, a fully automated ledger for the transaction of the information. Its benefits society as well as the economy of the country. The concept of cryptocurrency has made it possible to reduce the money supply inflationary. It reduces the pressure of the central banks for printing the money to improve the economic growth of the country. It started during the financial crisis year of 2008–2009. This concept has liberated the control of central authority and brought distributed transactions to the community.

The Block Chain technology is nowadays able to store the complete information of each transaction between the supplier, buyer, and ownership, and it also preserves the identity and its assets files in the proper proof way. It provides a network type of decentralized environment.

As a result of the discussion, it can be concluded that Block Chain technology in upcoming years provides an excellent network of machine learning, big data, and the Internet of things (IoT). This technology is shifting our lifestyle like paradigm, which affects our belonging environments of markets. This technology brings satisfaction during the transaction of any things like assets and information, since the third-party transaction has been omitted with this renowned technology. It also benefits to GDP growth of the nation as well as building the marketplaces more transparent in dealing the communicational matters with smart contracts. It will develop fruitful ideas to the readers, decentralized in the industry-oriented as well as economic infrastructure.

In the future, Block Chain technology can be used for the medical units for keeping the electronic records of the health-related data. It also includes the future aspect of the health authority. It will deal with the affairs related to how external parties will be able to use the patient's records from the hospital without breaking patient privacy. It is suggested that further research work can be assisted to investigate the economic implications and the impact of non-Bitcoin Block Chain applications.

References

Ben Ayed, A., & Belhajji, M. A. (2018). The block chain technology. *International Journal of Hyperconnectivity and the Internet of Things, 1*(2), 1–11. https://doi.org/10.4018/ijhiot.2017070101.

Bruyn, A. S. (2017). *Block chain an introduction* (Research Paper). Available at: https://beta.vu.nl/nl/Images/werkstuk-bruyn_tcm235-862258.pdf.

Butum, L. C., Stan, S. O., & Gaitanaru, A. (2018). Tools used in eLearning and specific risks associated with this. Enabling students, professors and researchers with ICT skills. In *The 14th International Scientific Conference eLearning and Software for Education* (pp. 275–284), Bucharest, March 2018. https://doi.org/10.12753/2066-026x-18-253.

Casino, F., Dasaklis, T. K., & Patsakis, C. (2019). A systematic literature review of block chain-based applications: Current status, classification and open issues. *Telematics and Informatics, 36,* 55–81. https://doi.org/10.1016/j.tele.2018.11.006.

Chain, M. (2013). *Markov chain Monte Carlo: Innovations and applications. Dynamical systems with applications using MATLAB*. https://doi.org/10.1017/CBO9781107415324.004.

Chandra, C., & Grabis, J. (2016). *Supply chain configuration*. https://doi.org/10.1007/978-1-4939-3557-4.

Essaf, F., et al. (2019). Privacy protection issues in block chain technology. *International Journal of Computer Science and Information Security (IJCSIS),* 124–131.

Foroglou, G., & Tsilidou, A. L. (2015). Further applications of the block chain. In *12th Student Conference on Managerial Science and Technology* (pp. 1–8), Athens, May 2015. https://doi.org/10.13140/rg.2.1.2350.8568.

Hacker, P., et al. (2019). *Regulating block chain: Techno-social and legal challenges—An introduction* (pp. 1–39).

Hübnet, R. (2007). *Strategic supply chain management in process industries*. In: *Lecture Notes in Economics and Mathematical Systems*.

Jani, S. (2019, July). *The emergence of block chain technology & its adoption in India* (pp. 1–12). https://doi.org/10.13140/rg.2.2.30997.58087.

Kesharwani, S. (2018). Block chain a peer-to-peer network: A holistic study from research firm to corporate house. *Global Journal of Enterprise Information System, 10*(2), 1–12. https://doi.org/10.18311/gjeis/2018/23394.

Konstantinidis, I., Siaminos, G., Timplalexis, C., Zervas, P., Peristeras, V., & Decker, S. (2018). Block chain for business applications. In *Business Information Systems*. Springer International Publishing. https://doi.org/10.1007/978-3-319-93931-5.

Kossow, N., & Dykes, V. (2018, January). *Linkages between block chain technology and corruption issues*. https://doi.org/10.1111/j.1445-5994.2006.01027.x.

Madavi, D. (2008). A comprehensive study on block chain technology. *International Research Journal of Engineering and Technology,* 1765–1770. Available at: http://www.mendeley.com/research/comprehensive-study-BlockChain-technology.

Minis, I., et al. (2011). *Supply chain optimization, design, and management: Advances and intelligent methods*.

Rantala, J. (2019). Block chain as a medium for transindividual collective. *Culture, Theory and Critique, 60*(3–4), 250–263. https://doi.org/10.1080/14735784.2019.1694213.

Sarmah, S. S. (2018). Understanding block chain technology. *Computer Science and Engineering, 8*(2), 23–29. https://doi.org/10.5923/j.computer.20180802.02.

Shaikh, Z. A., & Lashari, I. A. (2017). Blockchain technology: The new internet. *International Journal of Management Sciences and Business Research, 6*(4), 167–177.

Sultan, K., Ruhi, U., & Lakhani, R. (2018). Conceptualizing block chains: Characteristics & applications. In *Proceedings of the 11th IADIS International Conference Information Systems 2018, IS 2018* (pp. 49–57).

Www.Cybrosys.Com Www.BlockChainexpert.Uk1. (n.d.). (pp. 1–97).

Zheng, Z., Xie, S., Dai, H.-N., & Wang, H. (2017). Block chain challenges and opportunities: A survey. *International Journal of Web and Grid Services, 14*(4), 1–24. https://doi.org/10.1504/IJWGS.2018.10016848.

Application

Inventory Modeling and Inventory Control Application

Sayantani Mondal, Aman Khatoon, and Swarup Poria

Abstract In this chapter, the basic motivation of inventory modeling will be discussed. Briefly description of the developmental history of dynamic inventory models will be given. Different inventory control strategies will be reviewed. Modern inventory control method via dynamical system approach will be illustrated. Stability of the models will be discussed. Finally, application areas and usefulness of the models will be reported from real-life inventory control point of view. In the conclusion section, summary of the chapter will be written. Moreover, the areas for future research will be discussed.

Keywords Mathematical preliminaries · Economic Order Quantity · Deterministic inventory problems · Probabilistic models · Deterministic inventory control · Probabilistic inventory control · Conclusion · References

1 Introduction

The objective of this chapter is to present the basic concepts of economic order quantity (EOQ) models. Forecast of demand and inventory is valuable act for Trade Company. Demand and inventory modeling determined by access, so models are generated by the use of individual or separate advanced mathematical tools. Inventory management is one of the most challenging activities for manufacturing organizations. Aggarwal (1974) discussed inventory theory and applications. Braglia et al. (2013) proposed a dybamic approach for inventory management, which can be used for a definitely non stationary demand whose rate evolves both in mean and in variance. Su et al. (1999) presented a production inventory model for a variable market

S. Mondal (✉) · A. Khatoon · S. Poria
Department of Applied Mathematics, University of Calcutta, 92 A. P. C Road, Kolkata 700009, India
e-mail: sayantanimondal1412199@gmail.com

A. Khatoon
e-mail: aman.khatoon0@gmail.com

S. Poria
e-mail: swarup_p@yahoo.com

K. Kumar and J. P Davim (eds.), *Supply Chain Intelligence*, Management and Industrial Engineering, https://doi.org/10.1007/978-3-030-46425-7_7

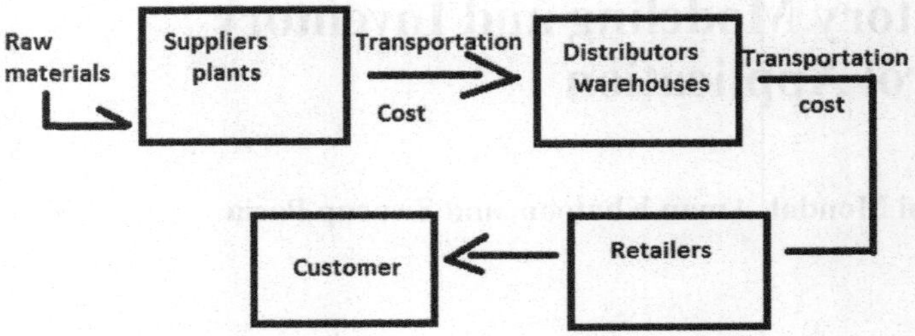

Fig. 1 Schematic diagram of inventory management

in which the production rate at any instant depends on the demand and the inventory level. EOQ model is the simplest and identical robust inventory model proposed by Ford Whitman Harris in 1913. Harris's model exhibits the substitution between inventory holding costs and ordering costs. Harvard University students accomplished the EOQ formula from Green (1915), later repeated by Copeland (1917). Next EOQ work had reported by Taft (1918b). Chandra et al. (1985) showed the effects of inflation and the time value of money on the optimal policies of the order-level system. Chen et al. (1997) offered a sensitivity analysis of the standard periodic review (s,S) inventory model. Gori et al. (2018) analyzed a heterogeneous agent model in which the fundamental exchange rate is endogenously determined by the real markets.

1.1 Mathematical Preliminaries

In this chapter, we shall discuss some basic items and mathematical preliminaries of the inventory problem which are relevant to the development of the next chapter. The flow chart of raw materials to the customer is shown in Fig. 1.

1.2 Inventory

An inventory problem can be defined as a stock of business articles to meet up the demand over a specific time period which implies relation among supplier and customer.

If there is an overstock the invested capital per unit time becomes large and if there is an overstock the invested capital per unit time becomes larger and there is an under stock the purchasing cost decreases but shortages occur, so the frequency of ordering increases. Inventory control is to arrange interminable production sales and customer service levels at the minimum cost.

1.3 Inventory Decision

In an inventory control situation, there are three basic questions to be answered. They are as follows:

1. How much to order? That is what is optimal of an item that should be ordered.
2. When should the order be placed?
3. How much safety stock should be kept? what quantity of items in excess requirements should be held as a buffer stock.

1.4 Inventory Cost

This costs accomplice alongside procurement, storage and management of inventory.

Ordering Cost: In the production process, required raw materials ordering cost is said as ordering cost. Ordering cost depends on workers' salaries, telephone calls, computer cost, transportation cost, analysis of raw materials, etc. A firm produced to its own product from exterior source then the cost of resetting for equipment production. The cost is expected as the cost per order or per setup.

Purchase Cost or production cost: Purchase cost is the procurement cost of unit items (or production). Important of the purchase price is when large production runs the process, these result in reduction of production cost or purchase cost per unit. For purchase "quality discounts" is specific quality, the ordered quantity may suitably be adjusted, so as to take advantage of these discounts.

Holding cost: holding cost usually denoted by C_1 or C_b per unit time.

The following components constitute the holding cost: Invested capital costs, Storage space cost, Handling costs, Recordkeeping costs, Deprecation, Deterioration and Obsolescence costs, Taxes and insurance costs, Salvage costs and selling price.

Setup cost: Set up cost is acquired to obtain material to provide another goods bunch. The variety of organizations this cost is independent.

Shortages or stock out: Shortages cost is a lack of inventory stock. This cost is included when customer fulfills their needs outside. Trouble of demand means business is not fulfilled by the company.

Salvages value: Salvages value is to measure value that benefits after fulfilling criteria.

1.5 Other Factors Involved in Inventory

Demand: Demand is a known number otherwise it will be completely unknown. Further, if it is known, it may be either fixed or variable per unit time. This introduces two types of inventory models. They are the following:

1. Deterministic Model: In this model, the demand is known and fixed.
2. Probabilistic Model: The model in which the demand is not known exactly which is assumed to be a random variable and is given by a known probability function is known as a probabilistic or stochastic model.

Replenishment: In a manufacturing system, the inventory of raw material, semi-finished goods, etc. may be required either to keep the warehouse for future need or to be used for immediate processing on arrival. The replenishment of such items may be instantaneous, constant or gradual, depending upon the lead time.

Order cycle: Time duration of placing one order to another order is known as order cycle. There are two types of order cycle: Continuous review, periodic review.

Lead time: Lead time is time interval within placing order and appearance of inventory. Level of inventory of an item depends upon the length of its lead time. It is divided into two parts one is Administrative lead time and other is Delivery lead time. Administrative lead time is the time for inception activity of store required. Delivery establishment for placing an order material is called as Delivery lead time.

Time horizon/Planning: Time horizon/planning is the time duration for inventory level until it is not composed. This time is either finite or infinite for the production of inventory.

Buffer (or safety) stock: Naturally, demand and lead time of product are not sure and it is not fixed. Therefore, demand deviation and supply deviation is not limited, it must be needed extra product/material to overcome the problem is called safety stock.

Number of items: Number of items is very important in inventory system. This system needs the extra stock to fulfill customer demands.

Maximum stock: Maximum stock level is inventory level that is not exceeded for inventory planning. Maximum stock level = Re-order level + Re-order quantity–(Minimum usage*minimum lead time).

The chart of inventory system is shown in Fig. 2.

1.6 EOQ (Economic Oder Quantity)

Company needs to control inventories. If it may grow up then the company faces suffer and similarly if it may be limited then the company may fail business. So, the optimal level of inventories can overcome this problem. Economic Order Quantity is a model to measure optimal volume that reduces total holding cost, ordering cost, shortages cost. These models are considered demand as a constant level and inventory are reduced constant amount until it reaches zero. No inventory lack is considered in this model. Two opposite costs can be shown by plotting them against the order size as shown in Fig. 3.

Inventory Models: Mathematical models or formula that can help us firm determining the EOQ and the frequency of ordering. Without interruption or delay of customer demand can play good services of firm. There are two types of inventory

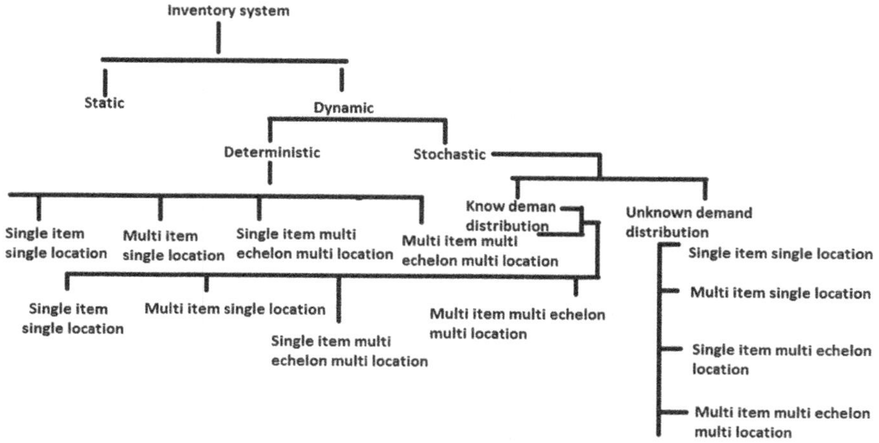

Fig. 2 Schematic diagram of basic inventory model

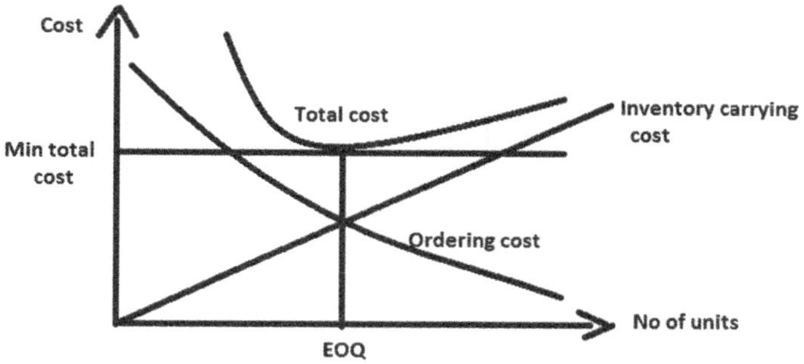

Fig. 3 Schematic diagram of EOQ model

models. Deterministic Inventory Models and Probabilistic or Stochastic inventory Models.

- **DETERMINISTIC INVENTORY PROBLEMES**

  Under this section, inventory models demand is considered immovable.

1.6.1 EOQ Model Without Shortage

In this model, we consider industrial description between optimum production quantity Q and R is Demand is uniform rate, Lead time is zero, the Production rate is infinite, Shortages are not allowed, C_1 is Holding cost, Set up cost is rupees C_s. Let each production cycle be made at fixed interval t and therefore, the quantity Q already presents in the beginning (when the business was started) should be.

$$Q = Rt,$$

where R is the demand rate since the stock in small time dt is Dtdt, the stock in total time t will be (Fig. 4).

$$= \int_0^t Rt\mathrm{d}t$$

$$= \frac{1}{2}Rt^2 = \frac{1}{2}Qt = \text{area of the inventory } \Delta\text{POA}, \ Rt = Q$$

Those cost of holding inventory per production run

$$= C_1 x \text{ area of } \Delta \text{ POA } = C_1\left(\frac{1}{2}Qt\right) = \frac{1}{2}C_1 Rt^2 \tag{1}$$

And the setup cost (i.e., production cost)

$$= \text{C}_\text{s} \text{ per production run for interval } t \tag{2}$$

Hence, summing up the cost in (1) and (2) and dividing by t, we get the total average cost given by,

$$C(t) = (1/2C_1 Rt^2 + C_s)/t$$

$$\text{Or, } C(t) = \frac{1}{2}C_1 Rt + \frac{C_s}{t} \tag{3}$$

is known as cost equation.

The condition for min. or max. of $C(t)$ in (3) and then equating it zero, we get

$$1/2C_1 R - C_s/t^2 = 0, \quad t = \sqrt{\frac{2C_s}{C_1 R}}, \tag{4}$$

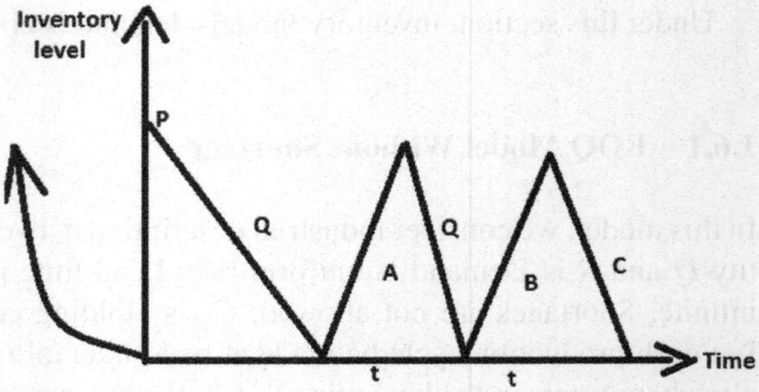

Fig. 4 EOQ model with no shortage and constant demand

Also differentiating $C(t)$ in (3) twice we get $(d^2C)/(dt^2) = (2C_S)/t^3 > 0$ for the final value of t given in Eq. (4).

Hence $C(t)$ is minimum for optimum time interval

$$t^* = \sqrt{\frac{2C_S}{C_1 R}} \tag{5}$$

And the optimum quantity to be produced (or ordered) at each level t^* is given by:

$$Q^* = DTt^* = D\sqrt{\frac{2C_S}{C_1 R}} = \sqrt{\frac{2C_S R}{C_1}} \tag{6}$$

Which is called the optimal lot size formula. This will result in minimum cost from Eq. (3) of the value

$$C_{\min} = \frac{1}{2} RC_1 \sqrt{\frac{2C_S}{RC_1}} + C_S \sqrt{\frac{RC_1}{2C_S}} = \sqrt{2C_1 RC_S} \tag{7}$$

Now if we consider the total time period be one year and this total time T be derived into n time interval of t then $nt = 1$ year and D be the total number of unit produced per year clearly $nQ = D$ Characteristic of model 1

1. Optimum under the orders placed per year

$$n^* = \frac{D^*}{Q^*} = \sqrt{\frac{DC_1}{2C_s}}$$

2. Optimum length of time between orders

$t^* = \frac{T}{n^*} = T\sqrt{\frac{2C_S}{Dc_1}}$ if $T = 1$ year

3. Minimum total annual inventory cost

$$TC^* = \frac{1}{2} Q^* C_1 + \frac{DC_S}{Q^*} = \sqrt{2C_S C_1}$$

1.6.2 The EOQ Model with no Shortage and Different Rates of Demands Different Cycle

In this model, D (demand) define over T (time) instead of demand rate being constant for each production cycle (Fig. 5).

Fig. 5 EOQ model with no shortage and variable demand

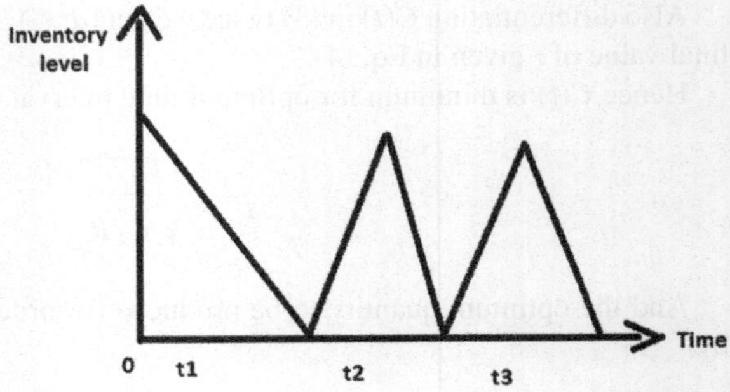

Let q be the fixed quantity produced in each production cycle. Since D define over T, the number n of production cycle will be given by $n = \frac{D}{Q}$
Also, let the vital period,

$$T = t_1 + t_2 + \cdots + t_n$$

It is obvious that fixed quantity Q, produced at the beginning of interval t_1, is supposed with a uniform rate of demand in interval t_1 only Then, obviously, the carrying cost for which the period T will be:

$$\left(\frac{1}{2}Qt_1\right)C_1 + \left(\frac{1}{2}Qt_2\right)C_1 + \cdots + \left(\frac{1}{2}Qt_n\right)C_1$$
$$= \left(\frac{1}{2}QC_1\right)(t_1 + t_2 + \cdots + t_n) = \frac{1}{2}C_1QT \quad (1)$$

$$\text{And the set up cost will be} = \frac{D}{Q}C_s \quad (2)$$

Thus we obtain the cost equation for period T as

$$C(Q) = \frac{1}{2}C_1QT + \frac{D}{Q}C_s \quad (3)$$

For optimum cost $\mathrm{d}C/\mathrm{d}Q = 1/2C_1T - C_s/Q^2D = 0$

$$Q = \sqrt{\frac{2C_s\left(\frac{D}{T}\right)}{C_1}} \quad (4)$$

Also $\frac{\mathrm{d}^2C}{\mathrm{d}Q^2} = \frac{2C_s}{Q^3}D > 0$ for value to Q at (4)
So, Optimum lot size is given by:
$Q^* = \sqrt{\frac{2C_s\left(\frac{D}{T}\right)}{C_1}}$ which minimizes the cost $C(Q)$ given by Eq. (3).
Substituting the value of Q^* in Eq. (3) we get,

$$C_{\min} = \frac{1}{2}(C_1 T\sqrt{\frac{2C_s(\frac{D}{T})}{C_1}} + C_s D\sqrt{\frac{C_1}{2C_s(\frac{D}{T})}}) = \sqrt{2C_1 C_s DT}$$

Hence minimum average cost will be

$$= \frac{\sqrt{2C_1 C_s DT}}{T} = \sqrt{2C_1 C_s\left(\frac{D}{T}\right)}$$

1.6.3 The EOQ Model with the Finite Rate of Replacement

Here assume as same as in model 1, except that of instantaneous replacement. Here we consider each production length t separates t_1 and t_2 inventory buildup for $(K - r)$ units, unit time t_1, $K > r$.

No replacement during t_2 where the inventory is decreasing factor rate r per unit of time. Where k is the production rate is finite and r is the demand rate also finite.

The diagram is given in Fig. 6.

Here total quality Q is produced over a period t_1 which is defined by the production rate K, since the inventory does not pile up in one shot but rather continuously over a time period and is also consumed simultaneously the average inventory level would be determined not only by the lot size Q, but also affected by the production rate K, depletion rate r.

Since t_1 produce Q at a rate K, we have $Q = Kt$,

Or, $t_1 = Q/K$.

During the production period, inventory accumulations at the rate of $(k - r)$ units,

So, average inventory $= \frac{1}{2}t_1(K - r) = \frac{1}{2}Q\left(1 - \frac{r}{k}\right)$.

Now with C_1 as the holding cost, the total annual holding cost $\frac{1}{2}QC_1\left(1 - \frac{r}{k}\right)$.

And the annual ordering cost is given by: $C_s \times \frac{D}{Q}$.

So, the total inventory cost is $C(Q) = \frac{1}{2}QC_1\left(1 - \frac{r}{k}\right) + C_s \times \frac{D}{Q}$.

Now for min or max $\frac{dC}{dQ} = 0$.

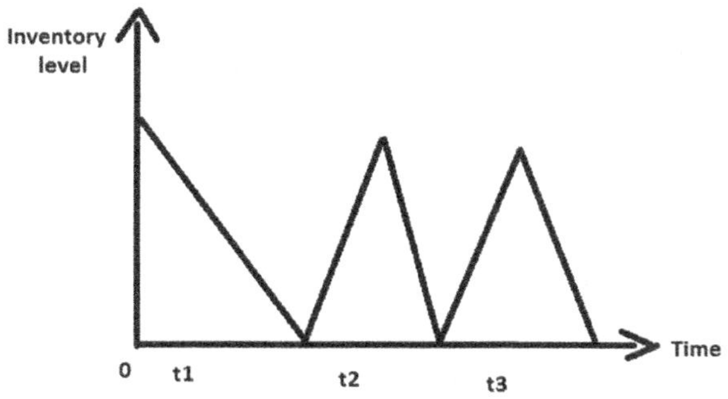

Fig. 6 EOQ model with finite rate of replacement

So, $Q = \sqrt{\frac{2DC_s}{C_1}}\sqrt{\frac{K}{K-r}}$.
And $\frac{d^2C}{dQ^2} = \frac{2C_sD}{Q^3} > 0$ for above value at Q.
C is minimum for these values of Q.
Hence $Q^* = \sqrt{\frac{2DC_s}{C_1(1-\frac{r}{K})}}$ optimal lot size formula,
$t_1^* = \frac{Q^*}{k}$ Optimal length of each lot size production.
$n^* = \frac{D}{Q^*}$ Optimal number of each production run year.
$C = \frac{D}{Q^*} + \frac{1}{2}Q\left(1 - \frac{r}{k}\right)$ minimum production inventory cost.

1.6.4 The EOQ Model with Shortage

In a business concern, if shortage occurs then these can be classified into the following two categories: (a) As soon as the desired units of a certain commodity arrive in inventory the back orders satisfied. (b) Shortage is lost sale. In the first case, the demand of the customer is met at the beginning of the new production run, whereas in the second case the customer move to some other firm to fulfill his requirements. We shall consider those types of problems of shortages where backorders are entertained.

Production is instantaneous: We assume that the production is instantaneous. Let the total time period be 1 year and let it is divided into n equal parts of interval t. Then $nt = 1$. Again each time interval t separates t_1 and t_2, such that $t = t_1 + t_2$. During the interval, t_1 times are drawn and in the next t_1 order for items are being accumulated but not filled. Then, the run size Q can be divided, into two parts Q_1 and Q_2, such that $Q = Q_1 + Q_2$. Q_1 Denotes the amount of inventory and Q_2 denotes the amount of demand. This inventory can be illustrated graphically in Fig. 7.

Now total inventory over the time period t is $\frac{1}{2}Q_1t_1$.
Average inventory holding cost $= \frac{\frac{1}{2}Q_1t_1C_1}{t}$.
Similarly, total amount of shortage overtime period $t = \frac{1}{2}Q_2t_2$.
Annual average shortage cost $= \frac{\frac{1}{2}Q_2t_2C_2}{t}$.

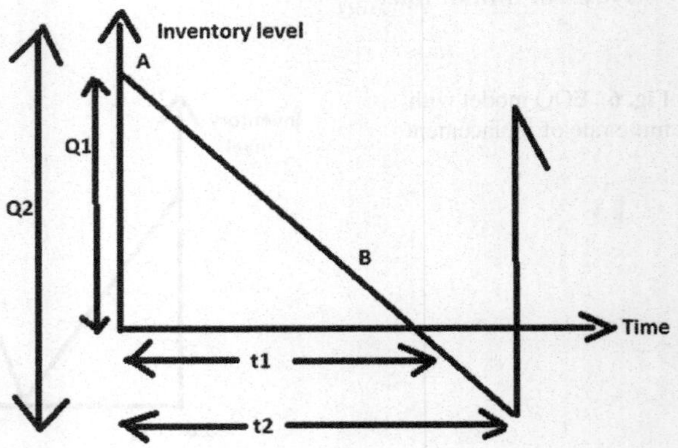

Fig. 7 EOQ model with shortage

An annual set up cost $= nC_s = \left(\frac{D}{Q}\right)C_s$.
So, total annual average cost is given by

$$C = \left[\frac{\frac{1}{2}Q_1 t_1 C_1 + \frac{1}{2}Q_2 t_2 C_2}{t} + \left(\frac{D}{Q}\right)C_s\right]$$

Now using the relationship for similar triangle ΔOAB and ΔBCD we get $\left[\frac{t_1}{Q_1} = \frac{t_2}{Q_2} = \frac{t}{Q}\right]$

$$t_1 = \frac{Q_1}{Q}t \text{ and } t_2 = \frac{Q_2}{Q}t$$

Making use of these values we get,
$C = \frac{1}{2}\left(\frac{Q_1^2}{Q}\right) + \frac{1}{2}C_2\left\{\frac{(Q-Q_1)^2}{Q}\right\} + C_s\left(\frac{D}{Q}\right)$ since $Q_2 = Q - Q_1$ for determining the optimum values of Q_1 and Q_1. So as to optimize C,
We have $\frac{\partial c}{\partial Q_1} = 0$

$$Q_1 = \frac{C_2 Q}{C_1 + C_2}$$

and $\frac{\partial C}{\partial Q} = 0$

$$Q = \sqrt{\frac{2C_s D + C_1 Q_1^2}{Q_1} + Q_1^2}$$

Now $\frac{\partial^2 C}{\partial Q^2} > 0$, $\frac{\partial^2 C}{\partial Q_1^2} > 0$ for those values of Q and Q_1
Optimum quantities are given by

$$Q^* = \sqrt{\frac{2C_s D}{C_1}}\sqrt{\frac{(C_1 + C_2)}{C_2}}$$

$$Q_1^* = \left(\frac{C_2}{C_1 + C_2}\right)Q^* = \sqrt{\frac{C_2}{C_1 + C_2}}\sqrt{\frac{2C_s D}{C_1}}$$

Now time between receipts of order

$$\frac{Q^*}{D_1} = t^*$$

Total optimum inventory cost

$$C^* = \sqrt{\frac{C_2}{C_1 + C_2}}\sqrt{2C_s DC_1}$$

Recorder level

$$Q_2^* = Q^* - Q_1^* = Q^*(1 - C_2(C_1 + C_2))$$

Remarks

(1) If $C_1 > 0, C_2 \to \infty$, shortage is restricted. So, $Q_1^* = Q^* = \sqrt{\frac{2C_sD}{C_1}}$, Q^* usage for inventory

(2) If $C_1 \to 0, C_2 \to \infty$, inventory is restricted. So, $Q_1^* = 0, Q^* = \sqrt{\frac{2C_sD}{C_s}}$, Q^* usage only fill backorders.

1.6.5 The EOQ Model with the Finite Rate of Replenishment with Shortage

Here assumed as same as in fundamental EOQ, except that the rate of replenishment of inventory is K units of quantity. Let us consider production range t divides two parts t_1 and t_2 then further subdivided another two t_{11}, t_{12}, t_{21} and t_{22} where,

- Inventory buildup at time t_{11},
- There is no replacement at time t_{12} at r/time.
- Shortages are developed at time t_{21} at r/time.
- Shortages are stuffed at time t_{22}.

Graphical representation of time situation o\is in Fig. 8.

From the above figure, we see that at the end of t_{11}, the level of inventory is Q_1 and at the end of period t_{12}, inventory becomes nil. New shortages start and suppose that the shortages buildup of quantity Q_2 up to time t_{21} and these filled up during time t_{22}.

Total inventory overtime period $t = \frac{1}{2}Q_1t_1$.

Total amount of shortage over the time period $t = \frac{1}{2}Q_2t_2$

Set up cost for one unit of time $= \frac{C_s}{(t_1+t_2)}$.

Now $t_1 + t_2 = Q/r$.

So, the total cost of one unit of time is given by,

$$C = C_1\frac{\left(\frac{1}{2}Q_1t_1\right)}{t} + C_2\frac{\left(\frac{1}{2}Q_2t_2\right)}{t} + \frac{rC_s}{Q}$$

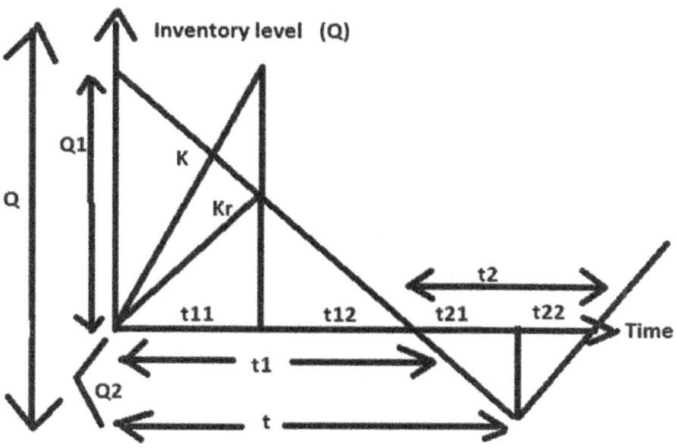

Fig. 8 EOQ model with finite rate of replacement and shortage

$$\frac{1}{2Q}\frac{K}{(K-r)}\left[C_1\left(\left(\frac{(K-r)}{K}Q-Q_2\right)^2+C_2Q_2^2\right)\right]+\frac{rC_s}{Q}$$

The optimum quantities are thus given by,

$$Q^* = \sqrt{\frac{2C_s(C_1+C_2)K_r}{C_1C_2(K-r)}}$$

$$Q_2^* = \sqrt{\frac{2C_1C_2(K-r)}{(C_1+C_2)KC_2}}$$

$$t^* = \sqrt{\frac{2(C_1+C_2)KC_2}{C_1C_2r(K-r)}}$$

$$Q_1^* = \sqrt{\frac{2C_1C_2r(K-r)}{(C_1+C_2)KC_1}}$$

$$C^* = \sqrt{\frac{2C_1C_2C_sr(K-r)}{(C_1+C_2)KC_1}}$$

- **Probabilistic or Stochastic Models**: This model demand is unknown and probability distribution of demand is known. The demand is a random variable, it may be either discrete or continuous. This model has two cases
  - Static demand models (single period models)
  - Dynamic demand models (multi-period with variable lead time models). In static demand models, there are three cases.
  - Instantaneous demand without setup cost models.
  - Continuous demand without setup cost models.
  - Instantaneous demand with setup cost models.

Here we will not give theoretical illustrations of these models.

Production inventory model with deteriorating elements and shortages

In this portion, we deal with continuous production, inventory model for deteriorating items with shortages. We are assuming the demand rate and production rate are constants. For the continuous distribution of deteriorating items, it follows the exponential distribution.

We are taking some numerical examples on the basis of our daily exchanges and further doing sensitivity analysis.

Notations and modeling assumption

Now we are taking the following notations as demand rate (a), p ($>a$) production rate, holding cost (C_1), C_2 shortage, C_3 deteriorated cost, inventory cost C, lead time is zero, durational time (T), exponential distribution $g(t)$ (θ deterioration amount; constant fraction θ ($0 < \theta \ll 1$)). For simplicity taken that no replacement at deteriorated component are added inventory.

Production starts when $t = 0$ and end with $t = t_1$. In the interval $[0, t_1]$ production rate is p and the demand rate is a ($<p$). Stock gain Q_1 when $t = t_1$. At time $t = t_2$, the stocks are ended and shortages occur, next, we accumulated Q when $t = t_3$. Production creates over p when $t = t_3$ and backlog is cleared at time $t = T$. Then cycle repeats itself after T (Fig. 9).

Following the above assumption, chart of the model is given below:

Mathematical Model Analysis

Consider $Q(t)$ at t ($0 \le t \le T$). Then

$$\frac{\mathrm{d}Q(t)}{\mathrm{d}t} + \theta Q(t) = p - a, \quad 0 \le t \le t_1 \tag{1}$$

Fig. 9 EOQ model with deteriorating elements and shortage

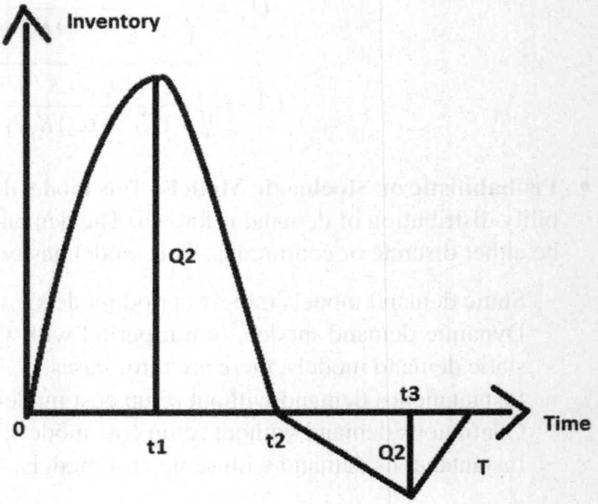

$$\frac{dQ(t)}{dt} + \theta Q(t) = -a, \quad t_1 \le t \le t_2 \tag{2}$$

$$\frac{dQ(t)}{dt} = -a, \quad t_2 \le t \le t_3 \tag{3}$$

$$\frac{dQ(t)}{dt} = p - a, \quad t_3 \le t \le T \tag{4}$$

And

$$Q(0) = 0,\ Q(t_1) = Q_1,\ Q(t_2) = 0,\ Q(t_3) = -Q_2,\ Q(T) = 0 \tag{5}$$

So solution of the above eqn

$$Q(t) = \frac{1}{\theta}(p-a)\left(1 - e^{-\theta t}\right), \quad 0 \le t \le t_1 \tag{6}$$

$$= -\frac{a}{\theta} + \left(Q_1 + \frac{a}{\theta}\right)e^{\theta(t_1 - t)}, \quad t_1 \le t \le t_2 \tag{7}$$

$$= a(t_2 - t),\ t_2 \le t \le t_3 \tag{8}$$

$$= (p-a)(t - t_3) - Q_2,\ t_3 \le t \le T \tag{9}$$

From (5) and (6), we get

$$Q_1 = Q(t_1) = \frac{1}{\theta}(p-a)\left(1 - e^{-\theta t_1}\right),$$

$$e^{\theta t_1} = \left[1 - \frac{\theta Q_1}{(p-a)}\right]^{-1}$$

$$t_1 = \frac{1}{\theta}\log\left[1 + \left\{\frac{\theta Q_1}{(p-a)} + \frac{Q_1^2\theta^2}{(p-a)^2}\right\}\right] \tag{10a}$$

$$= \frac{Q_1}{(p-a)} + \frac{\theta Q_1^2}{2(p-a)^2} \tag{10b}$$

(ignoring larger order θ, $0 \le \theta \le 1$)
Again From (5) and (7), we get

$$0 = Q(t_2) = -\frac{a}{\theta} + \left(Q_1 + \frac{a}{\theta}\right)e^{\theta(t_1 - t_2)}$$

$$t_2 - t_1 = \frac{1}{\theta}\log\left(1 + \frac{\theta Q_1}{a}\right) \tag{11}$$

$$t_2 = \frac{1}{\theta}\log\left[\left(1+\frac{\theta Q_1}{a}\right)\left\{1+\frac{\theta Q_1}{(p-a)}+\frac{\theta^2 Q_1^2}{(p-a)^2}\right\}\right] \quad \text{(using(10))} \tag{12}$$

Putting $(t_3) = -Q_2$, in Eq. (8),

$$a(t_2 - t_3) = -Q_2$$

$$t_3 = \frac{Q_2}{a} + t_2 \tag{13}$$

$$t_3 = \frac{Q_2}{a} + \frac{1}{\theta}\log\left[\left(1+\frac{\theta Q_1}{a}\right)\left\{1+\frac{\theta Q_1}{(p-a)}+\frac{\theta^2 Q_1^2}{(p-a)^2}\right\}\right] \tag{14}$$

Putting the condition $Q(T) = 0$ in Eq. (9),

$$(p-a)(T-t_3) = Q_2 \tag{15}$$

So, deterioration at [0, T]

$$\begin{aligned}
&= \{(p-a)t_1 - Q_1\} + \{Q_1 - a(t_2 - t_1)\} \\
&= \left[\frac{(p-a)}{\theta}\log\left\{1+\frac{\theta Q_1}{(p-a)}+\frac{\theta^2 Q_1^2}{(p-a)^2}\right\} - Q_1\right] \\
&\quad + \left[Q_1 - \frac{a}{\theta}\log\left(1+\frac{\theta Q_1}{a}\right)\right) \\
&= \frac{p}{\theta}\log\left\{1+\frac{\theta Q_1}{(p-a)}+\frac{\theta^2 Q_1^2}{(p-a)^2}\right\} \\
&\quad - \frac{a}{\theta}\log\left[\left\{1+\frac{\theta Q_1}{(p-a)}+\frac{\theta^2 Q_1^2}{(p-a)^2}\right\}\left\{1+\frac{\theta Q_1}{a}\right\}\right] \\
&= \frac{p}{\theta}\left\{\frac{\theta Q_1}{(p-a)}+\frac{\theta^2 Q_1^2}{(p-a)^2}-\frac{\theta^2 Q_1^2}{2(p-a)^2}\right\} \\
&\quad - \frac{a}{\theta}\left\{\frac{\theta Q_1}{(p-a)}+\frac{\theta^2 Q_1^2}{(p-a)^2}+\frac{\theta Q_1}{a}+\frac{\theta^2 Q_1^2}{a(p-a)}-\frac{\theta^2 Q_1^2 p^2}{2a^2(p-a)^2}\right\}
\end{aligned}$$

(Neglecting higher powers of θ)

$$= \frac{\theta Q_1^2 p}{2a(p-a)} \tag{16}$$

Deterioration cost over [0, T]

$$= \frac{c_3 \theta p Q_1^2}{2a(p-a)} \tag{17}$$

Shortage cost over $[0, T]$

$$c_2 \int_{t_2}^{T} \{-Q(t)\}\mathrm{d}t$$

$$= \left[\int_{t_2}^{t_3} a(t_2 - t)\mathrm{d}t + \int_{t_3}^{T} \{(p-a)(t-t_3) - Q_2\}\mathrm{d}t \right]$$

(from (8) and (9))

$$= \frac{C_2 Q_2^2 p}{2a(p-a)} \quad \text{(from (13) and (15))} \tag{18}$$

Carrying cost over $[0, T]$

$$= c_1 \int_0^{t_2} Q(t)\mathrm{d}t = c_1 \left[\int_0^{t_1} \frac{(p-a)}{\theta}(1 - e^{-\theta t})\mathrm{d}t \right.$$

$$\left. + \int_{t_1}^{t_2} \left\{ -\frac{a}{\theta} + \left(Q_1 + \frac{a}{\theta}\right) \mathrm{e}^{\theta(t_1 - t)} \right\} \mathrm{d}t \right] \tag{19}$$

Now,

$$\int_0^{t_1} \frac{(p-a)}{\theta}(1 - \mathrm{e}^{\theta t})\mathrm{d}t$$

$$= \frac{(p-a)}{2} t_1^2 \left(1 - \frac{\theta t_1}{3}\right) \quad \text{(ignoring higher order } \theta)$$

$$= \frac{Q_1^2}{2(p-a)} + \frac{\theta Q_1^3}{3(p-a)^2}$$

(from (10) and ignoring higher order θ) (20)

$$\int_{t_2}^{t_2} \left\{ -\frac{a}{\theta} + \left(Q_1 + \frac{a}{\theta}\right) \mathrm{e}^{\theta(t_1 - t)} \right\} \mathrm{d}t = \frac{a}{\theta}(t_1 - t_2)$$

$$+ \left(Q_1 + \frac{a}{\theta}\right) \frac{1}{\theta} \{1 - \mathrm{e}^{-\theta(t_2 - t_1)}\} = \frac{a}{\theta} . \frac{1}{\theta} \log\left(1 + \frac{\theta Q_1}{a}\right)^{-1}$$

$$+ \left(Q_1 + \frac{a}{\theta}\right) \frac{1}{\theta} \left\{ 1 - \left(1 + \frac{\theta Q_1}{a}\right)^{-1} \right\} \text{(by (11))}$$

$$= \frac{a}{\theta^2}\log\left\{1-\left(\frac{\theta Q_1}{a}-\frac{\theta^2 Q_1^2}{a^2}+\frac{\theta^3 Q_1^3}{a^3}\right)\right\}$$

$$+\left(Q_1+\frac{a}{\theta}\right)\frac{1}{\theta}\left\{1-\left(1-\frac{\theta Q_1}{a}+\frac{\theta^2 Q_1^2}{a^2}-\frac{\theta^3 Q_1^3}{a^3}\right)\right\}$$

$$= \frac{a}{\theta^2}\left\{-\frac{\theta Q_1}{a}+\frac{\theta^2 Q_1^2}{a^2}-\frac{\theta^3 Q_1^3}{a^3}-\frac{1}{2}\left(\frac{\theta^2 Q_1^2}{a^2}-\frac{\theta^3 Q_1^3}{a^3}\right)\right\}$$

$$+\left(Q_1+\frac{a}{\theta}\right)\left(\frac{Q_1}{a}-\frac{\theta Q_1^2}{a^2}+\frac{\theta^2 Q_1^3}{a^3}\right)$$

$$= \frac{Q_1^2}{2a}\text{(ignoring higher order }\theta) \tag{21}$$

So, carrying cost at [0, *T]*

$$= C_1\left\{\frac{Q_1^2}{2(p-a)}+\frac{\theta Q_1^3}{3(p-a)^2}+\frac{Q_1^2}{2a}\right\} = C_1\left\{\frac{Q_1^2 p}{2a(p-a)}+\frac{\theta Q_1^3}{3(p-a)^2}\right\} \tag{22}$$

So, total inventory cost (from (17), (18) and (22))

$$\begin{aligned}&= c(Q_1, Q_2)\\ &= \frac{c_1}{T}\left\{\frac{Q_1^2 p}{2a(p-a)}+\frac{\theta Q_1^3}{3(p-a)^2}\right\}\\ &+\frac{c_2 p Q_2^2}{2aT(p-a)}+\frac{c_3\theta p Q_1^2}{2aT(p-a)}\end{aligned} \tag{23}$$

And using (14), (15)

$$Q_2 = \frac{aT(p-a)}{p}-Q_1-\frac{\theta}{2}Q_1^2\frac{(2a-p)}{a(p-a)} \tag{24}$$

Using (23), (24) we get

$$\begin{aligned}&= c(Q_1) = \frac{c_1}{T}\left\{\frac{Q_1^2 p}{2a(p-a)}+\frac{\theta Q_1^3}{3(p-a)^2}\right\}\\ &+\frac{c_2 p}{2aT(p-a)}\left\{\frac{aT(p-a)}{p}-Q_1-\frac{\theta}{2}Q_1^2\frac{(2a-p)}{a(p-a)}\right\}^2\\ &+\frac{c_3\theta p Q_1^2}{2aT(p-a)},\end{aligned} \tag{25}$$

where $c(Q_1)$ is average cost function.

1.6.6 An EOQ Model for Deteriorating Items with Stock-Dependent Time-Varying Demand

Here EOQ models are used deteriorating items. In this model, demand is stock-dependent and time-dependent.

Notation and assumptions: In this model, $f(t) = (a + bt)\ [Q(t)]^{\beta}$ in $(0, T)$, $Q(t)$ = inventory at t which imply $0 \leq t \leq T$, a, b, β are positive constant $0 \leq \beta \leq 1$. Here β implies shape parameter, a = demand at $t = 0$, b = demand increases. h = holding cost, A = replenishment cost, C = item unit cost, T is length of cycle and θ $(0 < \theta \ll 1)$ is constant fraction.

Mathematical model analysis: Here uniform replenishment rate starts with inventory level q. At time T, inventory level reaches 0. Then the differential equation of $Q(t)$ is given by

$$\frac{dQ(t)}{dt} + \theta Q(t) = -(a + bt)[Q(t)]^{\beta}, \quad (a > 0, b > 0, 0 < \beta < 1) \tag{1}$$

When $Q(0) = q$, $Q(T) = 0$.
Then

$$[Q(t)]^{1-\beta} = q^{1-\beta}\exp(-\theta(1-\beta)t) - (1-\beta)\exp(-\theta(1-\beta)t)\int_0^t (a+bt)\exp(\theta(1-\beta)t)dt \tag{2}$$

Putting $Q(t) = 0$, we get

$$q^{(1-\beta)} = \left\{\frac{(a+bT)}{\theta} - \frac{b}{\theta^2(1-\beta)}\right\}\exp(\theta(1-\beta)T) - \left(\frac{a}{\theta} - \frac{b}{\theta^2(1-\beta)}\right)$$

Now from (2)

$$\begin{aligned}[Q(t)]^{(1-\beta)} &= \frac{1}{\theta}\left(a - \frac{b}{\theta(1-\beta)}\right)(\exp(\theta(1-\beta)(T-t)) - 1) \\ &+ \frac{b}{\theta}(T\ \exp(\theta(1-\beta)(T-t)) - t) \\ &= (1-\beta)(T-t)[(a+bT) + \frac{(T-t)}{2}\{\theta(1-\beta)(a+bT) - b\} \\ &+ \theta(1-\beta)\frac{(T-t)^2}{2}\{\theta(1-\beta)(a+bT) - b\} - \frac{\theta^2 b}{24}(1-\beta)(T-t)^3]\end{aligned} \tag{3}$$

Neglecting higher powers of θ.
Hence

$$Q(t) = \{(1-\beta)(T-t)(a+bT)\}^{\frac{1}{(1-\beta)}}$$
$$\left[1 + \frac{1}{(1-\beta)(a+bT)}\left\{\frac{(T-t)}{2}[\theta(1-\beta)(a+bT) - b\right]\right.$$
$$+ \frac{(T-t)^2}{6}(1-\beta)\{\theta^2(1-\beta)(a+bT) - b\theta\}$$
$$-\frac{\theta^2 b}{24}(1-\beta)(T-t)^3\Big\} + \frac{\beta}{2(1-\beta)^2(a+bT)^2}$$
$$\left[\frac{(T-t)^2}{4}\{\theta(1-\beta)(a+bT) - b\}^2 + \frac{\theta^2}{36}b^2(1-\beta)^2(T-t)^4\right.$$
$$+ \frac{(T-t)^3}{6}(1-\beta)\{b^2\theta - 2b\theta^2(1-\beta)(a+bT)\}$$
$$\left.\left.+\frac{\theta^2}{24}(T-t)^4 b^2(1-\beta)\right] - b\right] \tag{4}$$

inventory I_T is

$$I_T = \int_0^T Q(t)\mathrm{d}t = \{(1-\beta)(a+bT)\}^{\frac{1}{(1-\beta)}}\left[\frac{(1-\beta)}{(2-\beta)}T^{\frac{(2-\beta)}{(1-\beta)}}\right.$$
$$+ \frac{1}{(1-\beta)(a+bT)}\left\{\frac{(1-\beta)}{2(3-2\beta)}T^{\frac{(3-2\beta)}{(1-\beta)}}[\theta(1-\beta)(a+bT) - b]\right.$$
$$- \frac{\theta^2 b(1-\beta)^2}{24(5-4\beta)}T^{\frac{(5-4\beta)}{(1-\beta)}}$$
$$+ \frac{(1-\beta)^2}{6(4-3\beta)}\{\theta^2(1-\beta)(a+bT) - b\theta\}T^{\frac{(4-3\beta)}{(1-\beta)}}$$
$$+ \frac{\beta}{2(1-\beta)^2(a+bT)^2}\left[\frac{(1-\beta)}{4(4-3\beta)}T^{\frac{(4-3\beta)}{(1-\beta)}}\{b - \theta(1-\beta)(a+bT)\}^2\right.$$
$$+ \frac{\theta^2}{36(6-5\beta)}\{b^2(1-\beta)^3 T^{\frac{(6-5\beta)}{(1-\beta)}} + \frac{\theta^2 b^2(1-\beta)^2}{24(6-5\beta)}T^{\frac{(6-5\beta)}{(1-\beta)}}\Big\}$$
$$\left.\left.+\frac{(1-\beta)^2}{6(5-4\beta)}\{b^2\theta - 2b\theta^{2(1-\beta)(a+bT)}\}\right]\right] \tag{5}$$

$$D = q - \text{Total demand} = q - \int_0^T (a+bt)[Q(t)]^\beta \mathrm{d}t$$

$$= \left(\left\{\frac{(a+bT)}{\theta} - \frac{b}{\theta^2(1-\beta)}\right\}\exp(\theta(1-\beta)T) - \left(\frac{a}{\theta} - \frac{b}{\theta^2(1-\beta)}\right)\right)^{\frac{1}{(1-\beta)}}$$
$$- a[(1-\beta)(a+bT)]^{\frac{\beta}{(1-\beta)}}\left[(1-\beta)T^{\frac{1}{(1-\beta)}}\right.$$

$$+\frac{\beta}{(a+bT)}\left\{\frac{1}{2(2-\beta)}T^{\frac{(2-\beta)}{(1-\beta)}}[\theta(1-\beta)(a+bT)-b]\right.$$

$$+\frac{(1-\beta)}{6(3-2\beta)}\left[\theta^2(1-\beta)(a+bT)-b\theta\right]T^{\frac{(3-2\beta)}{(1-\beta)}}$$

$$\left.-\frac{\theta^2 b(1-\beta)}{24(4-3\beta)}T^{\frac{(4-3\beta)}{(1-\beta)}}\right\}+\frac{\beta(2\beta-1)}{2(1-\beta)(a+bT)^2}\left\{\frac{1}{4(3-2\beta)}T^{\frac{(3-2\beta)}{(1-\beta)}}\right.$$

$$[\theta(1-\beta)(a+bT)-b]^2+\frac{\theta^2 b^2(1-\beta)^2}{36(5-4\beta)}T^{\frac{(5-4\beta)}{(1-\beta)}}$$

$$+\frac{(1-\beta)}{6(4-3\beta)}\left[b^2\theta-2b\theta^2(1-\beta)(a+bT)\right]$$

$$\left.\left.+\frac{b^2\theta^2(1-\beta)}{24(5-4\beta)}T^{\frac{(5-4\beta)}{(1-\beta)}}\right\}\right]-b[(1-\beta)(a+bT)^{\frac{\beta}{(1-\beta)}}[\frac{(1-\beta)^2}{(2-\beta)}T^{\frac{(2-\beta)}{(1-\beta)}}$$

$$+\frac{\beta}{(a+bT)}\left\{-\frac{b\theta^2(1-\beta)^2}{24(4-3\beta(5-4\beta)}T^{\frac{(5-4\beta)}{(1-\beta)}}\right.$$

$$+\frac{(1-\beta)}{2(2-\beta)(3-2\beta)}[\theta(1-\beta)(a+bT)-b]T^{\frac{(3-2\beta)}{(1-\beta)}}$$

$$\left.+\frac{(1-\beta)^2}{6(3-2\beta)(4-3\beta)}\left[\theta^2(1-\beta)(a+bT)-b\theta\right]T^{\frac{(4-3\beta)}{(1-\beta)}}\right\}$$

$$+\frac{\beta(2\beta-1)}{2(1-\beta)(a+bT)^2}\left\{\frac{(1-\beta)}{4(3-2\beta)(4-3\beta)}[\theta(1-\beta)(a+bT)-b]^2T^{\frac{(4-3\beta)}{(1-\beta)}}\right.$$

$$+\frac{\theta^2 b^2(1-\beta)^3}{36(5-4\beta)(6-5\beta)}T^{\frac{(6-5\beta)}{(1-\beta)}}+\frac{\theta^2 b^2(1-\beta)^2}{24(5-4\beta)(6-5\beta)}T^{\frac{(6-5\beta)}{(1-\beta)}}$$

$$\left.\left.+\frac{(1-\beta)^2}{6(4-3\beta)(5-4\beta)}\left\{b^2\theta-2b\theta^2(1-\beta)(a+bT)\right\}T^{\frac{(5-4\beta)}{(1-\beta)}}\right\}\right] \tag{6}$$

where

$$[Q(t)]^\beta=[(1-\beta)(T-t)(a+bT)]^{\frac{\beta}{(1-\beta)}}$$

$$\left[1+\frac{\beta}{(1-\beta)(a+bT)}\left\{-\frac{\theta^2 b}{24}(1-\beta)(T-t)^3\right.\right.$$

$$+\frac{(T-t)}{2}[\theta(1-\beta)(a+bT)-b]+\frac{(T-t)^2}{6}$$

$$\left.(1-\beta)\left[\theta^2(1-\beta)(a+bT)-b\theta\right]\right\}$$

$$+\frac{\beta(2\beta-1)}{2(1-\beta)^2(a+bT)^2}\left\{\frac{(T-t)^2}{4}[\theta(1-\beta)(a+bT)-b]^2\right.$$

$$+\frac{\theta^2 b^2}{36}(1-\beta)^2(T-t)^4$$

$$+\frac{(1-\beta)(T-t)^3}{6}\left[b^2\theta-2b\theta^2(1-\beta)(a+bT)\right]$$

$$+\frac{\theta^2 b^2}{24}(1-\beta)(T-t)^4\Big\}\Big]\tag{7}$$

$$C(T)=\frac{1}{T}[A+CD+hI_T]\tag{8}$$

To regulate T^* minimizes C (T) inventory system. So, C (T) is

$$\frac{\mathrm{d}C(T)}{\mathrm{d}T}=0$$

which gives

$$=\frac{1}{T^2}[A+CD+hI_T]+\frac{1}{T}\left(h\frac{\mathrm{d}I_T}{\mathrm{d}T}+C\frac{\mathrm{d}D}{\mathrm{d}T}\right)=0\tag{9}$$

Provided T^* will satisfy

$$\frac{d^2C(T)}{\mathrm{d}T^2}>0\tag{10}$$

In the limit as $\theta\to 0, \beta\to 0, b\to 0$ then $T^*=\left(\frac{2A}{ah}\right)^{\frac{1}{2}}$, EOQ cycle time. Then, EOQ model has given below

$$q^*\to aT^*=\left(\frac{2\mathrm{aA}}{\mathrm{h}}\right)^{\frac{1}{2}}.$$

Again, *as* $\theta\to 0, \beta\to 0, b\to 0$

$$C(T^*)\to\frac{ahT^*}{2}+\frac{A}{T*}$$

Known as Wilson (1934) EOQ model. Also, it is determined as not possible and solves Eq. (9) to get value of T.

To solve (9) equation also we get T*, we get $C(T$*) from (8). If the solutions obtained, do not satisfy the sufficient condition (10). So, from Eq. (9) no appropriate optimal solution for the above parameter value.

2 Objectives of Inventory Control

Inventory control is a highly significant function not only because inventory constitutes a substantial portion of total assets of a affirm but also because it has to satisfy the following objectives:

1. To minimize the financial investment inventories.
2. To ensure that the value of the material consumed is minimum.
3. To maintain that timely records of inventories of all items and all items and to maintain the stocks within the desired limits.
4. To maintain timely action for replacement.
5. To meet demand fluctuations.

2.1 Deterministic Inventory Control

The deterministic inventory control based on that the parameter value and variables associated with an inventory are known and computing with the process certainty and replenishment lead time is constant and independent of the demand.

2.2 Probabilistic Inventory Control

In the probabilistic inventory control model, we assume that the average demands are reasonably constant over time, therefore, it is possible to describe a probability distribution of the demand, especially during replenishment lead time. This model is also called stochastic inventory control.

The flow chart of inventory control is shown in Fig. 10.

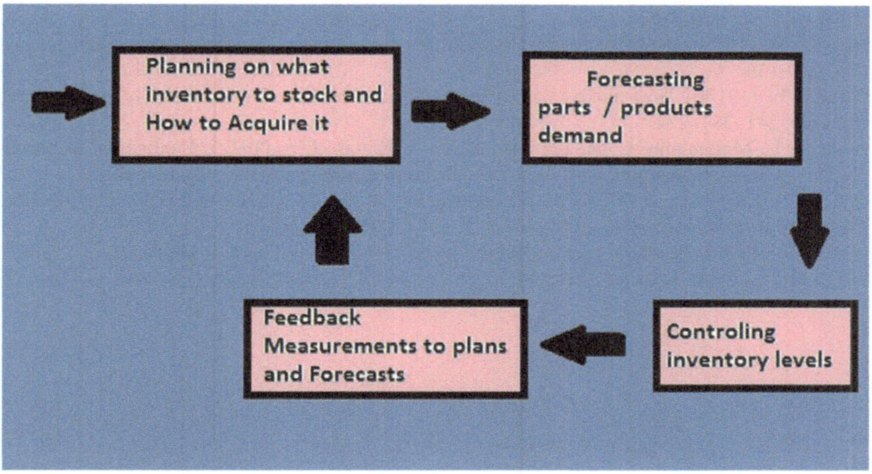

Fig. 10 Flow chart on inventory control

3 Conclusion

Nowadays we are totally dependent on online supermarkets, though the managers on artificial intelligence are worked on it. According to "Levin et al. (1972)" there *is good displayed in a supermarket will lead the customer to buy more.*

In this chapter, we have dealt with an inventory model for deteriorating items with a stock-dependent time-varying demand pattern. Though there is so much factor in the inventory demand, the retailer may take so much strategies, discount level high, stock clearance, showing high popularity, etc. Then there must be some problem is occurred such as space allocation space for items and investment money should increase inventory levels.

References

Aggarwal, S. C. (1974). A review of current inventory theory and it's applications. *International Journal of Production Research, 12,* 443–482.

Braglia, M., Gabbrieli, R., & Zammori, F. (2013). Stock diffusion theory: a dynamic model for inventory control. *International Journal of Production Research, 51,* 3018–3036.

Chandra, M. J., & Bahner, M. L. (1985). The effects of inflation and time value of money on some inventory systems. int. *J. Prod., 23,* 723–730.

Chen, F., & Zheng, Y. S. (1997). Sensitivity analysis of an (s, S) inventory model. *Operations Research letters, 21,* 19–23.

Copeland, M. T. (ed.) (1917). Business statistics. Harvard University Press, Cambridge, Mass.

Gori, M., & Ricchiuti, G. (2018). A dynamic exchange rate model with heterogeneous agents. *Journal of Evolutionary Economics, 28,* 399–415.

Green, J. B. (1915). The perpentual inventory in practical stores operation. *The Engineering Magazine, 48,* 879–888.

Harris, F. W. (1913a). How many parts to make at once. Factory. *The Magazine of management, 10,* 135–136.

Harris, F. W. (1913b). How much stock to keep on Hand. Factory. *The Magazine of management, 10,* 240–241, 281–284.

Levin, R. I., McLaughin, C. P., Lemone, R. P., & Kottas, J. F. (1972). Production/operations MManagement. Contemporary policy for managing operating system. McGraw-Hill.

Su, C. T., & Lin, C. W. (1999). A production inventory model for variable demand and production. *Yugoslav Journal of Operations Research, 9,* 197–206.

Taft, E. W. (1918b). The most economical production lot. *Iron Age, 101,* 1410–1412.

Wilson, R. H. (1934). A scientific routine for stock control. *Harvard Business Review, 13,*116–128.

Radial Data Envelopment Analysis Approach to Performance Measurement: Study on Indian Banking System

Preeti, Supriyo Roy, and Kaushik Kumar

Abstract Altering customer demands with augmented competitive pressures significantly change the operating environment of Indian banks. Being competitive, banks are continuously keeping themselves on a platform of performance as set by them. Traditional methods for evaluation of performance are not working properly due to different issues as floated in this sector. Considering all the above issues, performance measurement is very complex as well as critical in terms of banks long-term sustainability. This study adopts a non-traditional method of measuring performance via Data Envelopment Analysis; a frontier-based radial approach to measure performance between public and private sector banks for last five years (2015–2019). Using non-parametric test, this study also tests the hypothesis that '*there exist no statistical efficiency differences between the public and private sector banks*'. Analyses show a significant result that: *no bank is found to be overall technically efficient with all banks scoring less than one, and the private sector banks is 1.5% more technically efficient than public sector banks*. Most significantly, strong ownership effect is proofed to be significant on the performance of Indian banking system.

Keywords Operational efficiency · Radial model · Data envelopment analysis · Performance measurement · Indian banking system · Super-efficiency

1 Introduction

Beginning from early 1990s, Indian banking system underwent a striking transformation by introducing reforms relating to interest rate structure, cash reserve ratio, statutory reserve ratio, globalization and liberalization. Likewise, the present banking scenario is poised with challenges comprising of increasing stressed assets, slow growth of economy, demonetization, stringent capital adequacy norms, recapitalization of stressed public sector banks and consolidation of weaker banks (Reserve

Preeti (✉) · S. Roy
Department of Management, Birla Institute of Technology, Mesra, Ranchi, India
e-mail: preetimathotia@gmail.com

K. Kumar
Department of Mechanical Engineering, Birla Institute of Technology, Mesra, Ranchi, India

K. Kumar and J. P Davim (eds.), *Supply Chain Intelligence*, Management and Industrial Engineering, https://doi.org/10.1007/978-3-030-46425-7_8

Bank of India 2018). Altering customer demands with augmented competitive pressures significantly change the operating environment of banks. This calls for required changes in the bank's strategies to cope with sustainable growth of banks. Therefore, the banking sector is under constant pressure to deliver services efficiently and effectively. The situation still remains worrisome as banks performance is continuously affected due to plenty of mixed factors. Hence, measuring the performance of banks is a key concern for banking decision-makers.

Traditional ratio analysis methods were used for measuring the performance of banks. Demerits of this method in aggregating multiple input–output data in determining efficiency led to the usage of frontier-based approaches (Yeh 1996). Popular frontier approaches used by financial institutions for measuring efficiency are parametric and non-parametric approaches (Berger and Humphrey 1997). Stochastic Frontier Approach, a significant parametric approach, necessitates a functional form of input–output variables used and also accounts for random error. Due to this complication of specifying functional form, the non-parametric approaches like DEA are extensively used by industry experts and academicians (Zimková 2014). DEA, which uses linear programming approach, is used for decision making and benchmarking the DMU(s) that uses multiple inputs–outputs set. It constructs a Pareto optimal frontier from the set of efficient DMUs by enveloping all other inefficient DMUs (Charnes et al. 1978). Radial DEA model, though known as the basic DEA model, is still popular among academic researchers to measure performances across different sectors like manufacturing as well as services (Thanassoulis et al. 2011; Agarwal 2016; Kluczek 2017). Banking is one of the popular application fields of DEA as concluded in a study by Emrouznejad and Yang (2018).

Considering the state of Indian banking sector, the crux of the problem boils down mainly to performance management. In this sequel, contribution of the study may direct to the decision-makers in a more signified way by incorporating latest model-based analysis; followed by a structural output to infer performance issue(s) in between privately-owned and state-owned banks. Layout of the study is distributed into different sections. Section 2 presents review of literature that used DEA to determine the performance of banks. Brief explanation of DEA methodology is provided in Sect. 3. Empirical findings are illustrated in the following Sect. 4. The study concludes with Sect. 5.

2 Literature Review, Gaps and Objectives

Traditional performance measures like ratio and regression analysis are not considered suitable with multiple input and output of banks (Paradi et al. 2011). Against this backdrop, the researchers resorted to frontier-based approaches for performance measurement. Studies like Ferrier and Lovell (1990) and Bhattacharyya et al. (1997) used Stochastic Frontier Approach, a parametric approach to evaluate the performance of banking system. However, non-parametric approach is best fitted for measuring the performance of financial institutions over parametric approaches (Sherman and Gold 1985). DEA, a non-traditional approach is confirmed to be most popular after surveying 130 studies across 21 countries (Berger and Humphrey 1997). As a non-parametric approach, it is a frontier-based approach considers multiple input and output by comparing one DMU with another without any problem of statistical average (Berger and Humphrey 1997; Sathye 2015). After the pioneering work of Berger and Humphrey, a lot number of academic researchers have developed relevant DEA models to measure the performance of all service sectors.

2.1 Data Envelopment Analysis: Global Scenario

Studies confined to the United States and developed European countries are more compared to developing countries (Berger and Humphrey 1997). See and He (2015) witnessed an increase in bank efficiency within the period of 2003–2010 with a significant difference between joint-stock banks and state-owned banks. To overcome biases between efficiency scores, the study used double bootstrap DEA approach. Financial innovation in banking system can prove to increase competitiveness of banks. DEA was applied in the sample of 24 Romanian banks to study the contribution of using interest services to enhance operational performance (Stoica et al. 2015). The second stage analysis is increasingly used in DEA-based study to determine factors that best explain the efficiency of banks. Matousek and Nguyen (2016) developed bias-corrected radial DEA models to find efficiency increase for the period of 1999–2009. The study confirmed that large-sized banks are superior in performance than small and medium-sized banks. However, large network and older banks showed less efficiency compared to other banks. Post severe economic crises in 1997, Indonesia witnessed sharp regulatory changes. Defung et al. (2016) concluded a significantly positive relationship between technical efficiency score and regulatory reforms.

Malaysian banking sector exhibited improvement in efficiency compared to foreign banks over the period 1999–2008 by employing the DEA approach (Sufian et al. 2016). Basically, this study tested the significance of origin on the performance differences of banks. Non-interest income is a significant portion of income for Ghanaian banks. Two-step DEA approach is mostly implemented to confirm that exclusion of this income would lead to under-estimation of banks performance (Alhassan and Tetteh 2017). Moreover, the performance of Malaysian banks might also change depending on banks nature and ownership. Azad et al. (2017) studied the sample of 43 Malaysian commercial banks to confirm that bank ownership and nature greatly influences banks performance. Banks face different categories of risk that affect the performance of banks. Tan and Anchor (2017) found that Chinese banking industry is greatly affected by liquidity risk. Banking sector development, bank diversification, GDP growth rate, stock market development and inflation are among other significant factors that explain efficiency differences of the Chinese banking system.

Level of inefficiency and its main causes of Brazilian banking were studied for the period of 2012–2016 by using radial DEA models (Henriques et al. 2018). The study identified efficient and inefficient banks by estimating PTE and SE. Most recent study refers to the development of radial DEA model for Croatian banking industry during the period from 2006 to 2015 (Davidovic et al. 2019). In almost all the studies, variables like relative ownership structure, market size and origin of capital are studied to test the relationship with efficiency score.

2.2 *Data Envelopment Analysis: Indian Scenario*

Application of non-traditional approach for performance evaluation started slightly late in comparison to worldwide study. In India, Bhatia and Mahendru (2015) developed radial DEA to estimate the efficiency (profit, revenue and cost) of Indian state-owned banks for the study period 2008–2013. Improvement in the volume of non-performing loans and technology is suggested as important findings of the study. Radial DEA model is developed to find the performance of Indian banks and comment on the relationship between efficiency and profitability of banks (Singh and Kaur 2016). Kumar et al. (2016) evaluated both the productivity and efficiency status of Indian banks for the two significant periods- the post-reform period and the global financial crisis period.

Bootstrap DEA approach aims to eliminate the biases associated with efficiency scores. Sathye and Sathye (2017) developed a bootstrap DEA approach to confirm the significance of bank soundness, ATM intensity, ownership, risk and size on the performance of Indian banks. Capital adequacy means a minimum amount of capital any financial institution must have at any point of time. Capital adequacy score of different banks is compared using DEA model to confirm that public sector banks are incompetent in terms of capital adequacy (Samanta and Chakraborty 2018). On the other hand, PSB exceeds in terms of revenue efficiency in post-reform era (Bhatia and Mahendru 2018).

Efficiency and productivity results across bank's sizes and ownership are compared for the period between 2008 and 2017 using Malmquist indices DEA approach (Tamatam et al. 2019). During the study period from 2011 to 2015, using radial DEA model, the operating efficiency of state-owned banks is 62% and privately-owned banks is 47% (Mohapatra et al. 2019). This study further used a truncated regression model to investigate the relationship of banks' performance with intellectual capital. The cost, revenue and profit efficiency of Indian banks is studied for 2004–2013 using DEA approach. Finding indicates a constructive relation between cost and revenue efficiency. Also, the size of banks and its management are important influential factors for banking performance (Jayaraman and Srinivasan 2019).

2.3 Gaps and Objectives

Gauging deep into the literature on Indian banking system, it is inferred that studies show mixed profile of analysis using radial DEA models. This paper tends to delve into estimating the constituent of OTE by focusing on ownership differences. Further, this study utilizes the most recent data for Indian banking system to make useful insights on performance. Therefore, specific objective of the study is to *estimate* the OTE, PTE and SE for Indian bank (PSB and PB) for the study period for last five years. The study *assigns* ranks to PSB and PB on the basis of super-efficiency scores estimated using model developed by Andersen and Petersen (1993) and then *test* whether ownership holds importance in Indian domestic banking system.

3 Methodology and Data Collection

Identification of the problem of performance measurement in Indian banking system, followed by extensive literature review, signifies the above gaps and objective(s). Methodology for the above problem is developed and explained hereunder.

3.1 Data Envelopment Analysis

An operation research technique called DEA is popularly applied across different sectors to evaluate performance in terms of efficiency. DEA is an extended version of Farrell's (1957) single input–output mix case. DEA, a linear programming frontier-based approach used to find out the performance of relative DMU(s) with several inputs and outputs. The efficient DMUs determined as Pareto-efficient DMUs forms the frontier which envelops all other inefficient DMUs. The significant requirement of DEA suggests that number of DMUs must not be less than thrice the sum of input–output mix. Within DEA framework, a model exhibits either CRS or VRS. CRS model approach allows us to estimate the OTE of DMUs. Whereas, a VRS model approach measures the PTE. Relation between two radial models (CCR and BCC) is used to estimate SE. Inefficiency analysis includes three directions in which inefficient DMU(s) can be projected onto the efficient frontier. In *input-oriented,* an inefficient DMU condenses its input amounts by keeping the present level of outputs. In *output-oriented*, an inefficient DMU enhances its output levels by consuming the same level of inputs. Third direction includes input excesses (input slack) and output shortfalls (output slack), where both are optimized simultaneously. Analysis into different directions is worked upon by researchers to estimate the inefficiency of banks. This study adopts output-oriented radial models of DEA to analyze the OTE, PTE and SE.

3.2 Radial Models—CCR and BCC

Radial DEA models evaluate the radial or proportionate distance from the efficient frontier. CCR model, also know a CRS model; formulated by Charnes et al. (1978) determines the global or overall technical efficiencies of any DMU.

Notations for the model are clarified as follows:

$n =$ Total number of decision making units (DMUs)
$X = \{x_j\}$ set of inputs;
$Y = \{y_j\}$ set of outputs; $j = 1, 2, 3, \ldots, n.$ $x_j \geq 0$ and $y_j \geq 0$
$x_j =$ individual inputs for DMU_j
$y_j =$ individual outputs for DMU_j
$p =$ optimal efficiency value
$\lambda =$ positive intensity vector.

With the above notations, production possibility set defined in Eq. (1) for output-oriented CCR model is:

$$P_{CCR} = \{(x, y)|x \geq X\lambda, y \leq Y\lambda, \lambda \geq 0\} \quad (1)$$

This model evaluates the overall technical efficiency {OTE1 for DMU(x_j, y_j).

$$\text{Maximize } \mathbf{p} \quad (2)$$

$$\text{Subjected to: } x0 - X\lambda \geq 0;\ y0 - Y\lambda \geq 0;\ \lambda \geq 0 \quad (3)$$

The objective is to maximize Eq. (2) with constraints defined in Eq. (3). Here, p denotes objective optimal value.

CRS assumption of the CCR model is relaxed to obtain BCC model of DEA. It is a VRS model propounded by Banker et al. (1984). Convexity constraint is added to CCR mode to obtain equations for BCC model. This model measures PTE which excludes SE. BCC model calculates PTE or managerial efficiency which measures inefficiencies occurring due to short-term managerial underperformance, also referred to as managerial inefficiency or X-inefficiency. Mathematical equation is the same as CCR with an additional constraint: $\sum_{j=1}^{n} \lambda_j = 1$. Any DMU operating at sub-optimal scale is said to be scale inefficient DMU. Scale inefficiency occurs due to long-term inefficiency in size of operations. Therefore, OTE can be explained by PTE and SE. Relationship between efficiency values obtained by CCR and BCC model can be used to estimate SE. Mathematical formulation is as given below:

$$\text{OTE}_{\text{CCR}} = \text{PTE}_{\text{BCC}} * \text{SE}$$

3.3 Super-Efficiency Model

Andersen and Petersen (1993) formulated a model with the concept of super-efficiency. Basic theme behind super-efficiency model is that it excludes that DMU from the reference set which is under evaluation. Unlike the rest of DEA models, this model takes efficiency values to be greater than or equal to 1. Hence, the ability of this model to differentiate or rank the efficient DMU assists in post-DEA statistical analysis. Linear programming equation for super-efficiency model is to estimate the efficiency for DMU q, where DMU q is excluded from the reference frontier. The fully efficient DMU q in radial BCC model may have efficiency score greater 1. For total n DMUs, with its reference set having $n - 1$ DMUs. LP equation for super-efficiency model for DMU q is

Notations for variables used in the model development are as follows.

n = Total number of DMUs,
p^q_{CCR} = super-efficiency score of DMU q,
ϕ^{super}_q = overall technical efficiency score for DMU q,
s_i^- = Input slack,
s_u^+ = Output slack,
x_{ij} = amount of ith input used by DMU j,
y_{ij} = amount of ith output used by DMU j,
q = DMU excluded from the reference set,
ε = non-archimedean element.

With the notations as discussed above, die mathematical equation for input-oriented super efficiency CCR model is presented below:

$$\text{Minimise } p^{\mathrm{q}}_{\mathrm{CCR}} = \phi^{\mathrm{super}}_{\mathrm{q}} - \varepsilon\left(\sum_{i=1}^{m} s_i^- + \sum_{u=1}^{v} s_{\mathrm{u}}^+\right) \tag{4}$$

$$\text{Subject to: } \sum_{j=1, j\neq q}^{n} \lambda_j y_{uj} - s_u^+ = y_{uq}, \quad u = 1, 2, \ldots, v$$

$$\sum_{i=1, j\neq q}^{n} \lambda_{ij} - s_i^- = x_{iq}, \quad i = 1, 2, \ldots, m$$

$$s_i^-, s_u^+ \geq 0; \lambda_j(j \neq q) \geq 0, \quad j = 1, 2, \ldots, n \tag{5}$$

Above equations represent the super-efficiency input-oriented CCR model. However, the super-efficiency output-oriented radial model can be estimated by solving the reciprocal for optimal solutions obtained in the above model. $p^q_{BCC} = 1/p^q_{CCR}$, and λ,s_u^+,s_i^- adjusted by division with p^q_{CCR}, where p^q_{BCC} is super-efficiency score from BCC model.

3.4 Data Collection

According to past literature, the two widely used approaches for deciding multiple inputs and output to develop DEA model are '*the production approach*' and '*the intermediation approaches*' (Humphrey 1985). In a study, Berger and Humphrey (1997) confirmed that an approach that fits into estimating bank-level efficiency is intermediation, whereas production approach satisfies the demand of measuring branch-level performance. This approach views banks as intermediator that uses purchased funds as well as deposits to produce different bank assets. Present study selected multiple inputs (*Operating expenses, Assets and Deposits*) and multiple outputs (*Investment and Advances*) using the intermediation approach. The secondary data supporting this study are selected from recent annual publications by Indian Banks Association. Choice of variables is strictly in accordance with managerial objective to maximize banks earnings by increasing investments and advances and at the same time incurring input costs like operating expenses, deposits and advances. Literature reference for each selected variables is illustrated in Table 1.

4 Empirical Findings

Empirical analysis of performance result obtained by radial DEA models is discussed in this section. Differentiated scores of efficiency derived from super-efficiency

Table 1 Input and output variable with literature citations

Variable(s)	Literature support
Operating expenses	Sherman and Gold (1985), Bhattacharyya et al. (1997), Stoica et al. (2015), Azad et al. (2017)
Assets	Bhatia and Mahendru (2015), See and He (2015), Defung et al. (2016), Sufian and Kamarudin (2016), Alhassan and Tetteh (2017), Bhatia and Mahendru (2018), Henriques et al. (2018)
Deposits	See and He (2015), Bhatia and Mahendru (2015), Stoica et al. (2015), Singh and Kaur (2016), Kumar et al. (2016), Stewart et al. (2016), Defung et al. (2016), Sufian and Kamarudin (2016), Alhassan and Tetteh (2017), Azad et al. (2017), Bhatia and Mahendru (2018), Henriques et al. (2018)
Advances	Bhatia and Mahendru (2015), See and He (2015), Stewart et al. (2016), Defung et al. (2016), Sufian and Kamarudin (2016), Singh and Kaur (2016), Kumar et al. (2016), Alhassan and Tetteh (2017), Azad et al. (2017), Tan and Anchor (2017), Bhatia and Mahendru (2018), Henriques et al. (2018), Mohapatra et al. (2019)
Investments	Bhatia and Mahendru (2015), See and He (2015), Defung et al. (2016), Kumar et al. (2016), Sufian and Kamarudin (2016), Alhassan and Tetteh (2017), Bhatia and Mahendru (2018), Mohapatra et al. (20192019

output-oriented BCC model is used to rank the banks. Further, a non-parametric test is used to test the hypothesis pertaining to ownership difference publicly-owned and privately-owned banks. Analysis of data is carried out using computer software DEA SOLVER LV8. This is EXCEL based software providing easy access to a variety of DEA models for analysis.

4.1 Analysis of Efficiency Scores

As per Table 2, no bank is found to be OTE with score being less than 1. The number of inefficient banks is observed to rise over the study period from 30 (in 2015) to 34 (in 2019) with mean OTE declining over the period (shown in Table 3). Efficient frontier comprising of banks is constructed for a particular year which assists in measuring the radial (proportionate) distance of inefficient banks from the developed efficient frontier. The average OTE for Indian banking system for an entire five year period is 0.9122. It means that approximately 9% of inefficiency existing in banks can be removed by maximizing output with same level of inputs. IBL exhibited highest level of OTE score, whereas NB was found to be highly overall inefficient.

As already stated, OTE can be separated into two constituents called PTE and SE. Like OTE, PTE also explains the inefficiency associated with resource utilization. Mean PTE score for the study period is estimated to be 0.9434, with number of inefficient banks rising from 25 (in 2015) to 27 (in 2019). It means that around 6% of inefficiency (underperformance) in Indian banking system is explained due to lack of managerial efficiency is achieving the right amount of output with given level of inputs. The residual part of inefficiency in OTE is measured by SE. it implies that on average 4% of inefficiency in banks is because of sub-optimal scale size. On average, the study found that none of the banks are operating at optimal level over the period of the study. The finding concludes that overall inefficiency is mainly due to inefficiency in resource utilization and not scale inefficiency.

4.2 Return to Scale

Significant contribution of radial model of DEA is that it can determine the scale at which any DMU might be operating. Constant RTS exhibits the optimum scale size at which any DMU must operate. On the other hand, DMUs exhibiting increasing RTS lies beneath the optimal level of operation and must expand operations to achieve the optimum scale size. Decreasing RTS experiencing DMUs lies above the optimum scale size and must cut down its scale to achieve the optimality. Table 4 provides the number of banks experiencing increasing, constant and decreasing RTS for all the five year study period. In the year 2015, total of 17 (i.e. 42.5%) banks are experiencing IRS, which means they are operating below the scale size and requires expansion. In contrast, in the year 2018, 25 (i.e. 62.5%) banks are operating at decreasing RTS

Table 2 Overall, pure and scale efficiency scores for public and private sector banks

Bank No.	Bank name	OTE	PTC	SE	Bank No.	Bank name	OTE	PTC	SE
1	ABB	0.92916	0.94046	0.98798	21	CUB	0.99420	1.00000	0.99420
2	AB	0.98420	0.98506	0.99913	22	TMB	0.92088	0.93012	0.99007
3	BoB	0.98578	1.00000	0.98578	23	CSB	0.81384	0.90498	0.89929
4	BoI	0.92994	0.95382	0.97496	24	DB	0.76498	0.91544	0.83564
5	BoM	0.88318	0.89198	0.99013	25	FB	0.94928	0.95834	0.99055
6	CB	0.92832	0.97212	0.95494	26	JKB	0.89930	0.91774	0.97991
7	CBI	0.73854	0.76546	0.96483	27	KB	0.91622	0.92586	0.98959
8	CPB	0.93876	0.94118	0.99743	28	KYB	0.97722	0.98986	0.98723
9	DB	0.81854	0.83328	0.98231	29	LVB	0.94598	0.95644	0.98906
10	IB	0.96332	0.96446	0.99882	30	NB	0.63494	1.00000	0.63494
11	IOB	0.82626	0.84954	0.97260	31	RBL	0.93278	0.96664	0.96497
12	OBC	0.93248	0.94004	0.99196	32	SIB	0.96946	0.97330	0.99605
13	PSB	0.95318	0.95846	0.99449	33	ABL	0.96974	0.99000	0.97954
14	PNB	0.92128	0.97202	0.94780	34	DCB	0.97686	0.99576	0.98102
15	SB	0.97346	0.99320	0.98012	35	HDFC	0.95364	0.99012	0.96316
16	UCO	0.85402	0.85638	0.99724	36	ICICI	0.96416	1.00000	0.96416
17	UB	0.98304	0.99474	0.98824	37	IBL	0.99458	0.99756	0.99701
18	UBI	0.70036	0.71586	0.97835	38	KMB	0.96574	0.97914	0.98631
19	VB	0.94118	0.95420	0.98636	39	YES	0.96478	0.96592	0.99882
20	SBI	0.90634	1.00000	0.90634	40	IDBI	0.88946	0.89172	0.99747

Table 3 Overall descriptive statistics of efficiency scores

	2015	2016	2017	2018	2019
OTE scores					
No. of efficient DMUs	10	10	9	5	6
No. of inefficient DMUs	30	30	31	35	34
Average	0.9333	0.9347	0.9202	0.8971	0.8757
Max	1	1	1	1	1
Min	0.631	0.647	0.6197	0.6215	0.6134
Std. Dev.	0.0746	0.0794	0.1016	0.1034	0.1184
PTE scores					
No. of efficient DMUs	15	14	17	15	13
No. of inefficient DMUs	25	26	23	25	27
Average	0.9545	0.9573	0.9419	0.9414	0.9212
Max	1	1	1	1	1
Min	0.8013	0.7699	0.6301	0.6504	0.6382
Std. Dev.	0.0509	0.0564	0.0829	0.0832	0.1116

Table 4 Return to scale—public and private banks

RTS	2015	2016	2017	2018	2019
Increasing RTS	17	10	8	8	8
Constant RTS	20	17	10	7	11
Decreasing RTS	3	13	22	25	21

which means downsizing is required to achieve the optimal scale size. The year 2015 showed the highest number of banks with constant RTS, thus operating at optimal size.

4.3 *Ranking of Banks Based on Super-Efficiency Score*

Super-efficiency scores obtained via Andersen and Petersen Model for the banks are reported in Table 5. The topmost efficient banks, SBI, a PSB, dominate among sample for 40 banks with super-efficiency score equivalent to 2.9254. CUB, a PB holds the next place with super-efficiency score being totalling to 1.1434. Out to top 10 efficient banks, 6 banks belongs to private sector (CUB, ICICI, DCB, HDFC, YES and IBL). However, only 4 out of top 10 banks are public sector banks (BoB, UB and SB). CBI and UBI found to be holding second last and last position among the sample of 40 banks as per the super-efficiency scores.

Table 5 Super-efficiency ranking of banks

Bank name	Super-efficiency score	Ranking	Bank name		Ranking
SBI	2.9254	1	LVB	0.9741	21
CUB	1.1434	2	RBL	0.9666	22
ICICI	1.0891	3	BoI	0.9655	23
DCB	1.0680	4	PSB	0.9602	24
BoB	1.0678	5	VB	0.9587	25
HDFC	1.0450	6	CPB	0.9498	26
YES	1.0317	7	OBC	0.9460	27
IBL	1.0251	8	AB	0.9405	28
UB	1.0233	9	TMB	0.9301	29
SB	1.0185	10	KB	0.9259	30
ABL	1.0073	11	IDBI	0.9210	31
NB	1.0000	12	JKB	0.9177	32
AB	0.9981	13	DB	0.9154	33
KVB	0.9928	14	CSB	0.9067	34
PNB	0.9927	15	BoM	0.8945	35
CB	0.9854	16	UCO	0.8564	36
SIB	0.9803	17	IOB	0.8495	37
KMB	0.9791	18	DB	0.8333	38
FB	0.9789	19	CBI	0.7655	39
IB	0.9784	20	UBI	0.7159	40

4.4 *Efficiency Differences Between Ownerships*

Descriptive statistics of three types of efficiency measures—OTE, PTE and SE for both categories of banks are provided in Table 6. It is evident, OTE score for 20 PSB 0.904567, whereas for 20 PB is 0.919902. This explains that PB is 1.5% more technically efficient in maximizing outputs than the PSB. The managerial efficiency as indicated by PTE is also found to be higher in PB compared to PSB since values are 0.924113 and 0.96245 for PB and PSB, respectively. Further, the scale efficiency levels for PSB is better than PB as reflected by efficiency scores of 0.9790 and 0.9560, for PSB and PB, respectively.

Efforts are in the direction to evaluate the efficiency differences between PSB and PB by applying Wilcoxon Mann–Whitney test (non-parametric test) as it does not assume the underlying distribution of efficiency score to be normal. Comparison on the basis of median is made to check the hypothesis that there exist no statistical efficiency differences between the PSB and PB. As evident from Table 7, the test statistics prove that for all efficiency measures (OTE, PTE and SE), the null hypothesis is rejected at 5% level of significance. The hypothesis testing proves that there

Table 6 Sector-wise descriptive statistics of overall, pure and scale efficiency score

Statistics	Public sector banks	Private sector banks
OTE		
Mean	0.90457	0.91990
Median	0.92955	0.95146
SD	0.07975	0.08865
Max	0.98578	0.99458
Min	0.70036	0.63494
PTE		
Mean	0.92411	0.96245
Median	0.95401	0.96997
SD	0.08045	0.03571
Max	1.00000	1.00000
Min	0.71586	0.89172
SE		
Mean	0.97899	0.95595
Median	0.98607	0.98677
SD	0.02221	0.08507
Max	0.99913	0.99882
Min	0.90634	0.63494

Table 7 Hypothesis testing between public and private sector banks

Ho	$\text{Median}_{\text{Public}} = \text{Median}_{\text{Private}}$
OTE	0.0007
PTE	0.01596
SE	0.02191
Decision	Reject Ho

is significant difference between efficiency level of OTE, PTE and SE for Indian banking system. It means that there is a consequence of ownership structure on the performance of banks in Indian banking system. This means that PB enjoys certain advantages over PSB in terms of utilization of resources.

5 Conclusion

Due to globalization coupled with fierce competition, recent period is considered as challenges in line of increasing stressed assets, slow growth of economy, demonetization, stringent capital adequacy norms, recapitalization of stressed PSB and consolidation of weaker banks. This study delves into the period of last five years to compute the operational performance of state-owned and privately-owned banks by developing radial DEA model by using input variables (*Operating expenses, Assets and Deposits*) and output variables (*Investment and Advances*). It measures the OTE, PTE and SE by using CCR and BCC models. Additionally, the super-efficiency concept; as developed by Andersen and Petersen (1993) to assign ranks by differentiating between efficient banks. Most significant finding of these analyses is: '*no bank is found to be overall technically efficient with each bank scoring less than 1*'. The year 2015 showed the highest number of banks with constant RTS, thus operating at optimal size.

As per super-efficiency ranking, SBI, a public sector bank, dominates all the other banks in overall. Result show CBI and UBI are found to be holding second last and last position. Comparison between PSB and PB revealed that PB is 1.5% more technically efficient than PSB during the study period. Hypothesis testing using non-parametric test proved that: '*there is significant difference between efficiency level of OTE, PTE and SE PSB and PB; which signifies strong ownership effect on the performance of banks*'. Present study limits to study only PSB and PB, excluding foreign banks, which also exist; but with small numbers. Future research can be focused on including foreign banks into this analysis. Another area where the present study may be explored are in line of examining certain exogenous factors and their influence on performance.

Appendix

No.	Full name	Code	No.	Full name	Code	No.	Full name	Code
1	Allahabad Bank	ABB	18	Dhanlaxmi Bank Limited	DBL	35	Punjab National Bank	PNB
2	Andhra Bank	AB	19	Federal Bank Limited	FB	36	Pure Technical Efficiency	PTE
3	Axis Bank Limited	ABL	20	HDFC Bank Limited	HDFC	37	Public sector bank	PSB
4	Bank of Baroda	BoB	21	ICICI Bank Limited	ICICI	38	Private sector bank	PB
5	Bank of India	BoI	22	IDBI Bank Limited	IDBI	39	RBL Bank Limited	RBL
6	Bank of Maharashtra	BoM	23	Indian Bank	IB	40	Return to scale	RTS
7	Banker, Cooper and Rhodes	BCC	24	Indian Overseas Bank	IOB	41	Scale efficiency	SE

(continued)

(continued)

No.	Full name	Code	No.	Full name	Code	No.	Full name	Code
8	Canara Bank	CB	25	Indusind Bank Limited	IBL	42	South Indian Bank Limited	SIB
9	Catholic Syrian Bank Ltd.	CSB	26	Jammu & Kashmir Bank Ltd.	JKB	43	State Bank of India	SBI
10	Central Bank of India	CBI	27	Karnataka Bank Limited	KB	44	Syndicate Bank	SB
11	Charnes, Cooper and Rhodes	CCR	28	Karur Vysya Bank Ltd.	KVB	45	Tamilnad Mercantile Bank Ltd.	TMB
12	City Union Bank Ltd.	CUB	29	Kotak Mahindra Bank Limited	KMB	46	UCO Bank	UCO
13	Corporation Bank	CPB	30	Lakshmi Vilas Bank Ltd.	LVB	47	Union Bank of India	UB
14	Data Envelopment Analysis	DEA	31	Nainital Bank Limited	NB	48	United Bank of India	UBI
15	DCB Bank Limited	DCB	32	Oriental Bank of Commerce	OBC	49	Variable return to scale	VRS
16	Decision Making Unit	DMU	33	Overall Technical Efficiency	OTE	50	Vijaya Bank	VB
17	Dena Bank	DB	34	Punjab & Sind Bank	PS	51	YES Bank Limited	YES

References

Agarwal, S. (2016). DEA-neural networks approach to assess the performance of public transport sector of India. *Opsearch, 53*(2), 248–258.

Alhassan, A. L., & Tetteh, M. L. (2017). Non-interest income and bank efficiency in Ghana: A two-stage DEA bootstrapping approach. *Journal of African Business, 18*(1), 124–142.

Andersen, P., & Petersen, N. C. (1993). A procedure for ranking efficient units in data envelopment analysis. *Management Science, 39*(10), 1261–1264.

Azad, M. A. K., Munisamy, S., Masum, A. K. M., Saona, P., & Wanke, P. (2017). Bank efficiency in Malaysia: A use of malmquist meta-frontier analysis. *Eurasian Business Review, 7*(2), 287–311.

Banker, R. D., Charnes, A., & Cooper, W. W. (1984). Some models for estimating technical and scale inefficiencies in data envelopment analysis. *Management Science, 30*(9), 1078–1092.

Berger, A. N., & Humphrey, D. B. (1997). Efficiency of financial institutions: International survey and directions for future research. *European Journal of Operational Research, 98*(2), 175–212.

Bhatia, A., & Mahendru, M. (2015). Measurement and evaluation of cost, revenue and profit efficiency of public sector banks in India. *International Journal of Behavioural Accounting and Finance, 5*(3–4), 298–313.
Bhatia, A., & Mahendru, M. (2018). Measurement of revenue efficiency of scheduled commercial banks across ownership in India. *IUP Journal of Bank Management, 17*(2), 20–49.
Bhattacharyya, A., Lovell, C. K., & Sahay, P. (1997). The impact of liberalization on the productive efficiency of Indian commercial banks. *European Journal of Operational Research, 98*(2), 332–345.
Charnes, A., Cooper, W. W., & Rhodes, E. (1978). Measuring the efficiency of decision making units. *European Journal of Operational Research, 2,* 429–444.
Davidovic, M., Uzelac, O., & Zelenovic, V. (2019). Efficiency dynamics of the Croatian banking industry: DEA investigation. *Economic research-Ekonomska istraživanja, 32*(1), 33–49.
Defung, F., Salim, R., & Bloch, H. (2016). Has regulatory reform had any impact on bank efficiency in Indonesia? A two-stage analysis. *Applied Economics, 48*(52), 5060–5074.
Emrouznejad, A., & Yang, G. L. (2018). A survey and analysis of the first 40 years of scholarly literature in DEA: 1978–2016. *Socio-Economic Planning Sciences, 61,* 4–8.
Farrel, M. J. (1957). The measurement of productive efficiency. *Journal of the Royal Statistical Society, 120*(3), 253–290.
Ferrier, G. D., & Lovell, C. K. (1990). Measuring cost efficiency in banking: Econometric and linear programming evidence. *Journal of Econometrics, 46*(1–2), 229–245.
Henriques, I. C., Sobreiro, V. A., Kimura, H., & Mariano, E. B. (2018). Efficiency in the Brazilian banking system using data envelopment analysis. *Future Business Journal, 4*(2), 157–178.
Humphrey, D. B. (1985). Costs and scale economies in bank intermediation'. In R. C. Aspinwall & R. A. Eisenbeis (Eds.), *Handbook for banking strategy* (pp. 745–783). New York: Wiley.
Jayaraman, A. R., & Srinivasan, M. R. (2019). Determinants of Indian banks efficiency: A two-stage approach. *International Journal of Operational Research, 36*(2), 270–291.
Kluczek, A. (2017). Assessing measures of energy efficiency improvement opportunities in the industry. *LogForum, 13*(1). http://dx.doi.org/10.17270/J.LOG.2017.1.3.
Kumar, M., Charles, V., & Mishra, C. S. (2016). Evaluating the performance of Indian banking sector using DEA during post-reform and global financial crisis. *Journal of Business Economics and Management, 17*(1), 156–172.
Mohapatra, S., Jena, S. K., Mitra, A., & Tiwari, A. K. (2019). Intellectual capital and firm performance: Evidence from Indian banking sector. *Applied Economics, 51*(57), 6054–6067.
Paradi, J. C., Rouatt, S., & Zhu, H. (2011).Two-stage evaluation of bank branch efficiency using data envelopment analysis. *Omega, 39*(1), 99–109.
Reserve Bank of India. (2018). *Report on trend and progress of banking in India 2017–18.* Retrieved from https://www.rbi.org.in.
Samanta, S., & Chakraborty, T. (2018). An analysis of capital adequacy in select Indian commercial banks: A frontier approach. *IUP Journal of Management Research, 17*(2).
Sathye, M. (2015). Technical efficiency of large bank production in Asia and the Pacific. *Multinational Finance Journal, 9*(1), 1–22.
Sathye, S., & Sathye, M. (2017). Do ATMs increase technical efficiency of banks in a developing country? Evidence from Indian Banks. *Australian Accounting Review, 27*(1), 101–111.
See, K. F., & He, Y. (2015). Determinants of technical efficiency in Chinese banking: A double bootstrap data envelopment analysis approach. *Global Economic Review, 44*(3), 286–307.
Sherman, H. D., & Gold, F. (1985). Bank branch operating efficiency: Evaluation with data envelopment analysis. *Journal of Banking & Finance, 9*(2), 297–315.
Singh, R. I., & Kaur, S. (2016). Efficiency and profitability of public and private sector banks in India: Data envelopment analysis approach. *IUP Journal of Bank Management, 15*(1).
Stewart, C., Matousek, R., & Nguyen, T. N. (2016). Efficiency in the Vietnamese banking system: A DEA double bootstrap approach. *Research in International Business and Finance, 36*(1), 96–111.
Stoica, O., Mehdian, S., & Sargu, A. (2015). The impact of internet banking on the performance of Romanian banks: DEA and PCA approach. *Procedia Economics and Finance, 20*(2), 610–622.

Sufian, F., & Kamarudin, F. (2016). Determinants of efficiency in the Malaysian banking sector: Does bank origins matter? *Intellectual Economics, 10*(1), 38–54.

Tamatam, R., Dutta, P., Dutta, G., & Lessmann, S. (2019). Efficiency analysis of Indian banking industry over the period 2008–2017 using data envelopment analysis. *Benchmarking: An International Journal, 26*(8), 2417–2442.

Tan, Y., & Anchor, J. (2017). The impacts of risk-taking behaviour and competition on technical efficiency: Evidence from the Chinese banking industry. *Research in International Business and Finance, 41*(4), 90–104.

Thanassoulis, E., Kortelainen, M., Johnes, G., & Johnes, J. (2011). Costs and efficiency of higher education institutions in England: A DEA analysis. *The Journal of the Operational Research Society, 62*(7), 1282–1297.

Yeh, Q. J. (1996). The application of data envelopment analysis in conjunction with financial ratios for bank performance evaluation. *Journal of the Operational Research Society, 47*(8), 980–988.

Zimková, E. (2014). Technical efficiency and super-efficiency of the banking sector in Slovakia. *Procedia Economics and Finance, 12*, 780–787.

Index

K. Kumar and J. P Davim (eds.), *Supply Chain Intelligence*, Management and Industrial Engineering, https://doi.org/10.1007/978-3-030-46425-7

The manufacturer's authorised representative in the EU is Springer Nature Customer Service Centre GmbH, Europaplatz 3, 69115 Heidelberg, Germany. If you have any concerns regarding our products, please contact ProductSafety@springernature.com

Printed and bound by CPI Group (UK) Ltd, Croydon, CR0 4YY
30/09/2026
02225665-0014